DESIGN OF ELECTRONIC CIRCUITS AND COMPUTER AIDED DESIGN

DESIGN OF ELECTRONIC CIRCUITS AND COMPUTER AIDED DESIGN

M.M. SHAH

JOHN WILEY & SONS
New York Chichester Brisbane Toronto Singapore

First Published in 1993 by
WILEY EASTERN LIMITED
4835/25 Ansari Road, Daryaganj
New Delhi 110 002, India

Distributors:

Australia and New Zealand:
JACARANDA WILEY LIMITED
PO Box 1226, Milton Old 4064, Australia

Canada:
JOHN WILEY & SONS CANADA LIMITED
22 Worcester Road, Rexdale, Ontario, Canada

Europe and Africa:
JOHN WILEY & SONS LIMITED
Baffins Lane, Chichester, West Sussex, England

South East Asia:
JOHN WILEY & SONS (PTE) LIMITED
05-04, Block B, Union Industrial Building
37 Jalan Pemimpin, Singapore 2057

Africa and South Asia:
WILEY EASTERN LIMITED
4835/25 Ansari Road, Daryaganj
New Delhi 110 002, India

North and South America and rest of the World:
JOHN WILEY & SONS, INC.
605 Third Avenue, New York, NY 10158, USA

Library of Congress Cataloging-in-Publication Data

ISBN 0-470-21954-8 John Wiley & Sons, Inc.
ISBN 81-224-0472-3 Wiley Eastern Limited
Typeset by CompuGraphy, O-4 Sri Niwas Puri, New Delhi - 110 065
and printed at Baba Barkha Nath Printers, New Delhi, India.

PREFACE

Scores of books have been written and are available on the principles of semiconductor devices and their applications. However, rarely one comes across a book which expounds these principles in actual practice. An electronic circuit rigged up by an amateur electronic engineer, or for that matter, by a 'learned theoretician', will almost invariably refuse to behave the way it is designed to. It is a highly frustrating experience and is guaranteed to dishearten and depress even the most ardent enthusiast. The risk is far less, if proper care is taken during the designing stage of the circuit. At least, one knows what could be the possible cause of the failure.

Most of the material presented in this book is the culmination of over 27 years of teaching similar subjects to various grades of students like diploma-grade and fifth and seventh semester students of various disciplines, like Electronic Engineering, Computer Engineering, Instrumentation Engineering and Electrical Engineering. The book should be of immense help to them. Indeed, this book will provide valuable guidance to a spectrum of users, from hobbyists to professional engineers.

The book is divided in two parts.

Part 1 deals with rectification, and *dc* regulated supplies using diodes, transistors and SCRs and linear circuits like amplifiers, both voltage and power, making use of BJTs and FETs. First, a brief analysis of the circuits is given, followed by the design procedures for these circuits with the help of numerous examples.

Part 2 of the book is devoted to the use of the computer in the design of the above circuits, after briefly dwelling on the device modelling and network topology and its solution by the computer.

For the first part, the author has assumed a working knowledge of the mathematics involved, as also of rudimentary network practices and of the principles of the operation of the semiconductor devices.

The second part, the Computer Aided Design of Electronic Circuits, demands a little more knowledge including a brush with BASIC and PASCAL. Most of the problems given in the book are written in PASCAL.

I am indeed thankful and indebted to a lot of people for help during the preparation of this book. I would like to mention at least a few of them; Both my daughters, Shital and Sejal, the former, a computer engineer, for writing some of the programmes presented in the text, and the latter for typing some of the most difficult formulae in the text, which can be considered a typist's nightmare. From the rest, scores of my students deserve my thanks, as also Mr. Kumbhavdekar of M/s Wiley Eastern, without whose untiring reminders regarding the tight time-schedule, I am sure, I wouldn't have completed this book.

Suggestions for improvement of the text, both for its presentation and material content, are welcome. I shall be grateful if errors, if any, and different approaches to the problems, if possible and helpful, are brought to my notice.

M.M. SHAH

CONTENTS

PART II

COMPUTER AIDED DESIGN OF ELECTRONIC CIRCUITS

PART I

DESIGN OF ELECTRONIC CIRCUITS

1

POWER SUPPLIES

The term *DC* power supply, or voltage supply is normally associated with, and popularly means, a conversion from *AC* to *DC*, or from *DC* to *AC*. Here we shall restrict ourselves to the first, briefly analysing and then designing *AC* to *DC* converters. In a later chapter, we shall attempt to design voltage regulators. The methods employed for the conversion or rectification are analysed here under.

1.1 Half-Wave Rectifier Circuits

The circuit shown in Figure 1.1 is a half wave rectifier circuit. The diode converts the *ac* sinusoidal wave to that shown in the diagram, since the diode effectively acts as a short circuit when forward biased, and open circuit when reverse biased. We shall analyse the circuit with a numerical example.

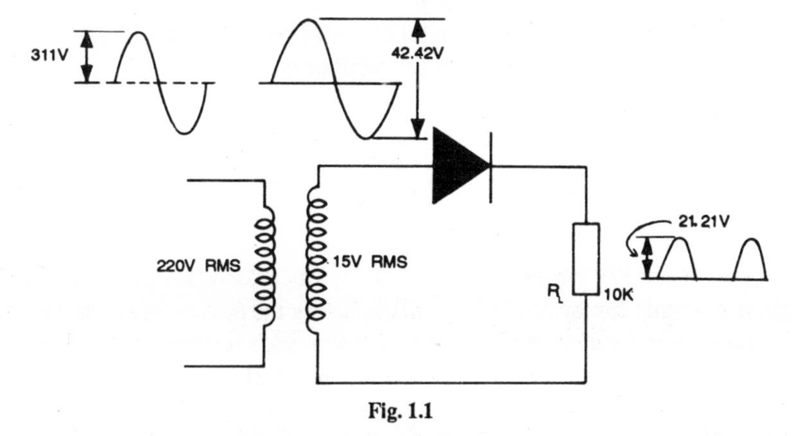

Fig. 1.1

Refer to the values given in the circuit diagram. For a 220 V to the primary of the transformer (i.e. $220 \times \sqrt{2} = 311$ Vp), the secondary of the transformer has 15 V rms, or 42.42 V p-p voltage. For this circuit, we have:

(a) Primary voltage peak $= 220 \times \sqrt{2} = 311$ V for a 220 V rms value.

(b) Secondary voltage *p-p* $= 2 \times 15 \times \sqrt{2} = 42.42$ V

(c) Peak voltage across the load = *p-p* voltage/2 = 21.21 V

(d) Since the load resistance is of 10K value, the peak current flowing will be $= V_L / R_L = 21.21/10\text{K} = 2.121$ mA

(e) V_{av} is the average value or the *dc* content of the voltage across the load and is given by

$$= \frac{1}{2\pi} \int_0^{\pi} V_p \sin \omega t \cdot d\,\omega t$$

$$= \frac{V_p}{2\pi} [-\cos \omega t]_0^{\pi} = \frac{V_p}{2\pi} \times 2 = \frac{V_p}{\pi}. \qquad (1.1)$$

In this case the average value = $V_p/3.1415 = 6.75$ V

(f) Average value of the current = $V_{av}/R_L = 6.75/10\text{K} = 0.675$ mA

(g) RMS voltage at the load resistance can be calculated as =

$$V_{rms} = \left[\frac{1}{2\pi} \int_0^{\pi} V_p^2 \cdot \sin^2 \omega t \cdot d\,\omega t \right]^{1/2} = V_p \left[\frac{1}{2\pi} \int_0^{\pi} \sin^2 \omega t \cdot d\,\omega \right]^{1/2}$$

$$= V_p \left[\frac{1}{2\pi} \int_0^{\pi} (1 - \cos 2\,\omega t)\, d\,\omega t \right] = \frac{V_p}{2} \qquad (1.2)$$

In this case the *rms* value = $V_p/2$

= peak voltage divided by a factor of 2 = 21.21/2 = 10.605 V

(h) The ripple factor for the half-wave rectifier is 1.21, and is arrived at by the formula:

$$\sqrt{\left(\frac{V_{rms}}{V_{DC}}\right)^2 - 1} = \sqrt{\left(\frac{V_p/2}{V_p/\pi}\right)^2 - 1}\ [(\pi/2)^2 - 1]^{1/2} = 1.21 \qquad (1.3)$$

i. The *PRV* or *PIV* is a parameter of a diode, which denotes a maximum reverse voltage that can be applied to a diode before the device breaks down. The peak reverse voltage that is applied across a diode is the peak of the negative half-cycle; in the example shown, it will be 21.21 V.

1.2 Full-Wave Rectifier Circuits Using Center-Tapped Transformer

The full-wave circuit shown in Fig. 1.2 employs a center-tapped transformer giving 15 - 0 - 15 V at the secondary. The center-tap is grounded. Effectively, this circuit can be visualised as having two half-wave rectifier circuits 180° out of phase with each other. Therefore, the amplitude of the wave remains the same, but double the number of half-waves are available.

(a) As in the case of the hw rectifier the p-p voltage available at the anode of each diode is = $2 \times \sqrt{2} \times 15 = 42.42$ V

It should be noted that there is a phase shift of 180° available between the two waveforms at the anodes of the diodes.

(b) The peak voltage available across the load resistance is the same as before, i.e. $\sqrt{2} \times 15 = 21.21$ volts.

(c) The average voltage available across the load resistance is

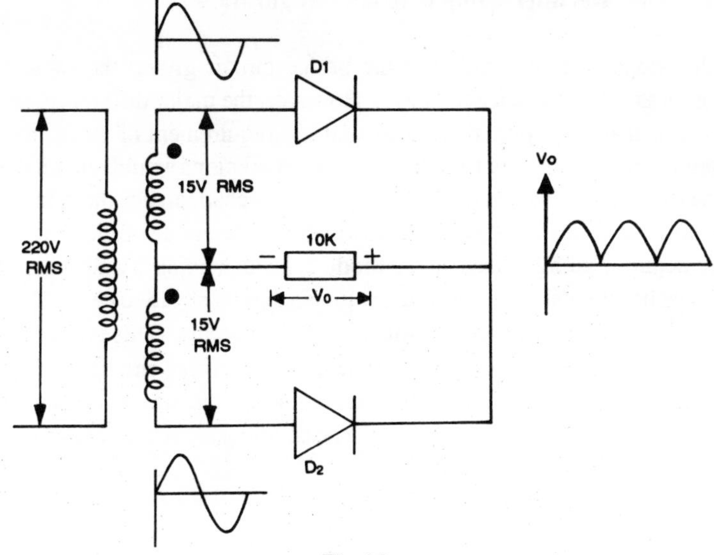

Fig. 1.2

$$= \frac{1}{\pi} \int_0^{\pi} V_p \sin \omega t \cdot d\,\omega t$$

$$= \frac{V_p}{\pi} [-\cos \omega t]_0^{\pi} \tag{1.4}$$

$= 2\, V_p/\pi = 13.5$ V, since double the number of half-waves are available.

(d) The average, or the *dc* value of the current flowing through R_L is given by $V_{av}/R_L = 1.35$ mA.

(e) RMS value of the voltage at the load resistance can be calculated as

$$= \sqrt{\frac{1}{\pi} \int_0^{\pi} V_p^2 \cdot \sin^2 \omega t \cdot d\,\omega t}$$

$$= \frac{V_p}{\sqrt{2}} \tag{1.5}$$

proceeding in the same way as in equation 1.2

V_{rms} = peak voltage divided by a factor of $\sqrt{2}$

= 15 volts.

(f) The ripple factor for the full-wave rectifier is given as 0.48, and is arrived at by the same formula as given for half wave rectifier.

1.3 Full-Wave Rectifier Using Bridge Configuration

The other popular configuration is the bridge circuit giving the same type of waveforms as the one discussed above. However, the major difference, as shown in Fig. 1.3 is that this circuit does away with the requirement of the center tapped transformer, but uses instead four diodes. The calculations would remain the same. Only the diode rating with reference to its *PIV* will change. For the center-tapped circuit, positive going voltage which is available at the anode of **diode 1** coincides with the negative going voltage at the **diode 2**, i.e. at a point in time when the +ve voltage reaches its peak, the −ve voltage also reaches its negative peak. This makes it 2 × (peak voltage) for the diode which is not conducting. This is not the case in bridge circuit, as can be readily seen. Hence, for the bridge circuit, the *PIV* rating of the diode has to be equal to the peak value of the waveform; and for the center-tapped circuit it should be twice the peak voltage. For the sake of reference, the formulae are given in Table 1.1.

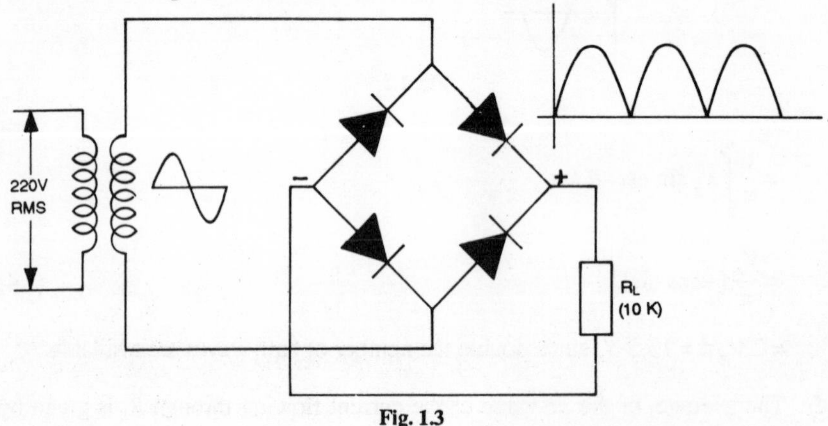

Fig. 1.3

There are two factors which need a little elaboration. They are as under.

1. Efficiency

If the entire circuit is considered as one unit, then the (total output)/(total input) factor can be considered as a measure of efficiency of rectification. Since the output is *dc* and the input is *ac*, we may write;

Efficiency = *dc* output-power delivered to the load/*ac* power input from the transformer secondary.

Hence, efficiency for the half-wave rectifier =

$$P_{dc}/P_{ac} = \frac{(I_{av})^2 \cdot R_L}{(I_{ac})^2 \cdot R_L} = \frac{(I_m/\pi)^2}{(I_m/2)^2} = \frac{4}{\pi^2} = 40.5\%$$

And the same for the full-wave rectifier $= \dfrac{(I_{av})^2 R_L}{(I_{ac})^2 R_L}$

$$= \frac{(2I_m/\pi)^2}{(I_m/\sqrt{2})^2} = \frac{8}{\pi^2} = 81.1\%$$

It should be noted that in calculating the above, we have neglected the forward resistance of the diode, which if considered, will make the efficiency to be little less than the values derived above.

2. Rating

On a similar basis it is useful to find a way so that rating of the transformer can be calculated, if the *dc* output power to be delivered to the load is known. For instance, for hw circuit to deliver a load current of I_{dc} at a voltage V_{dc}, i.e., a *dc* power of $P_{dc} = V_{dc} \times I_{dc}$, we will need a transformer which has

$$V_{rms} = V_p/\sqrt{2}, \text{ but since } V_{dc} = V_p/\pi$$

we have $V_{rms} = (\pi/\sqrt{2}) \times V_{dc}$.

Where the current in the winding flows for only one half cycle, the transformer will have to be considered as supplying current $I_{rms} = (\pi/2) \times I_{dc}$. The rating in volt-amp. of the secondary of the transformer

$$= V_{rms} \times I_{rms} \cdot = \left(\frac{\pi}{\sqrt{2}}\right) V_{dc} \times \left(\frac{\pi}{2}\right) \cdot I_{dc}$$

$$= 3.49\, P_{dc}.$$

Similarly, for the full-wave bridge rectifier circuit, this figure is given as

$$\text{Rating of the secondary } = V_{rms} \times I_{rms}$$

$$= \left(\frac{\pi}{\sqrt{2}}\right) \times V_{dc} \times \left(\frac{\pi}{\sqrt{2}}\right) \cdot I_{dc}$$

$$= 1.233\, P_{dc}.$$

For the center-tapped circuit this figure is 1.44.

This factor, often called, transformer utilisation factor, needs special consideration in half-wave rectifier circuits, since the current in the transformer winding flows during the same type of half cycles, creating a danger of the core getting saturated. Such transformers need special design consideration.

EXAMPLE 1.1

A full-wave bridge type circuit is designed with the transformer having a turns-ratio of 10:1, the primary voltage being 220 V. Analyse the circuit assuming the load resistance to be 100 ohms.

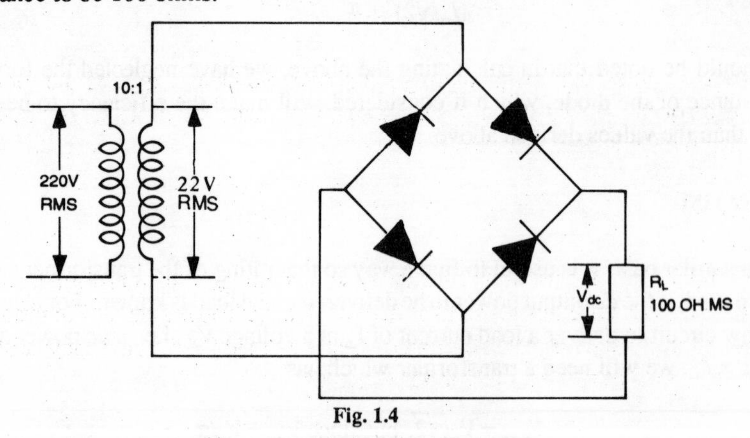

Fig. 1.4

(a) The *ac* rms voltage on the secondary of the transformer is = 220/10 = 22 V

(b) Average voltage = $V_{dc} = 2\ V_p/\pi = 19.8$ V

(c) $I_{dc} = V_{dc}/R_L = 19.8/100 = 198$ mA

(d) $P_{dc} = V_{dc} \times I_{dc} = 19.8 \times .198 = 3.92$ W

(e) *AC* power output needed to be delivered by the secondary of the transformer $= 1.23 \times 3.92 = 4.82$ VA.

(f) Assuming 85% efficiency for the transformer, the input to the primary of the transformer = 4.82/0.85 = 5.67 VA

g. The ripple factor $= \sqrt{\left(\frac{V_{rms}}{V_{DC}}\right)^2 - 1}$

$= 0.484.$

At this point, it is worthwhile to introduce yet one more variable, namely, the effect of the forward resistance of the diodes and the resistance of the transformer winding. If, in the above example, we assume a lumped resistance of, say, 3 ohms for the transformer and, say, 1 ohm per diode as the resistance at the point of operation of the diode, some of the values calculated will change slightly. We will consider the important ones.

Refer to Fig. 1.5.

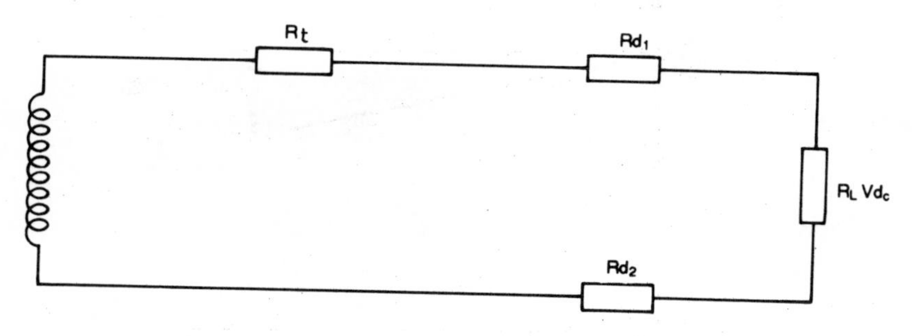

Fig. 1.5

The voltage available across the load resistance will be 2 V_P/π as calculated, if there is no load current flowing. However, as the load current increases, there will be greater voltage drop across the resistances and, hence, the output voltage available will go on decreasing, as is evident from the diagram and also from the equation

$$V_{dc} = (2V_p/\pi) - I_{dc}\,(R_t + 2R_d)$$

$$V_{dc} = \frac{(2\times\sqrt{2}\times 22)}{\pi} - \frac{V_{dc}}{R_L}(3+2)$$

$$= 19.8 - (3+2)\times V_{dc}/100$$

Therefore, $$V_{dc}\,(1 + 1/20) = 19.8, \text{ giving}$$

$$V_{dc} = 18.857$$

This introduces yet another factor, called the Regulation factor, which can be defined as $(V_{nL} - V_L)/V_L$

where V_{nL} = no-load output voltage
V_L = output voltage at load (normally full-load)

In the example given, Regulation factor = (19.8 − 18.857)/18.857
= 5.00 %

Frequently, the drop in the voltage, i.e., the difference in the full-load and the no-load voltages, is referred to as the no-load voltage instead of the full load voltage as used above. In this case, the regulation factor will be a slightly smaller figure, 4.76% instead of 5.00%.

The plot of the load-current versus the output voltage is given in Fig. 1.6. The drop in the characteristic is a measure of the internal resistance of the power-supply.

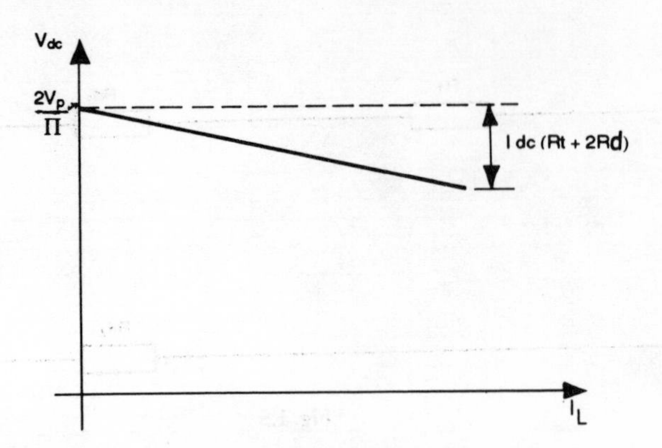

Fig. 1.6

Table 1.1

	Half-wave	*Full-wave (C.T.)*	*Bridge*
No of diodes	1	2	4
V_{dc} (no load)	V_p/π	$2\,V_p/\pi$	$2\,V_p/\pi$
Ave. current/diode	I_{dc}	$I_{dc}/2$	$I_{dc}/2$
Ripple factor	1.21	0.482	0.482
Transformer rating	3.49	1.745	1.233
Peak Inverse volt.	V_p	$2\,V_p$	V_p

2

FILTERS

A power supply must provide essentially a ripple free output voltage. The ripple in the rectified wave being very high, the factor being 48% in the full wave and 121% in the half-wave rectifier, majority of the applications which cannot tolerate this, will need an output which has been further processed. As it was stated earlier, the voltage wave-form available across the load resistance consists of two half-wave circuit outputs which are essentially 180° out of phase with each other, and the voltage, or the current equation can be given by

$$i = I_p \left[\frac{2}{\pi} - \frac{4}{\pi} \sum_{K=2,4,6,8} \frac{\cos \omega L}{(K+1)(K-1)} \right]$$

As can be seen, the fundamental frequency, ω, has been eliminated from the equation, and the lowest frequency that remains is 2ω, which is a second harmonic term. This clearly makes it convenient for use and design of a filter circuit. Such a circuit needs to be added between the rectifier and the load.

2.1 Inductor Filters

The operation of the inductor filter depends on its well known fundamental property to oppose any change of current passing through it. Without going into much technical details, we can conclude here that, since the *hw* wave-form is discontinuous, i.e. since the voltage or the current is absent at the output during the negative half-cycle of the input supply, the inductor circuit cannot function properly during this time, especially since the diode offers an almost open switch to the circuit.

For this reason an inductor filter is seldom used with half-wave circuit.

To analyse this filter for a full-wave, it will be necessary to write the equation as a Fourier series, which is

$$v = \frac{2V_p}{\pi} - \frac{4V_p}{\pi} \left[\frac{1}{3} \cos 2\omega t + \frac{1}{15} \cos 4\omega t + \frac{1}{35} \cos 6\omega t + \ldots \right] \tag{2.1}$$

The average, or the *dc* component is $= 2 \times V_p / \pi$, and can also be derived by the classical averaging method. Out of the remaining frequency components, only the lowest will be of significance, viz., frequency, $2 \times \omega$. The amplitude of all the other remaining components are small enough to be neglected.

$$\text{The current } i = \frac{2V_p}{\pi R_L} - \frac{4V_p}{3\pi} \cdot \frac{\cos(2\omega t - \phi)}{\sqrt{R_L^2 + 4\omega^2 L^2}} \tag{2.2}$$

where $\phi = \tan^{-1}(2\omega L/R_L)$

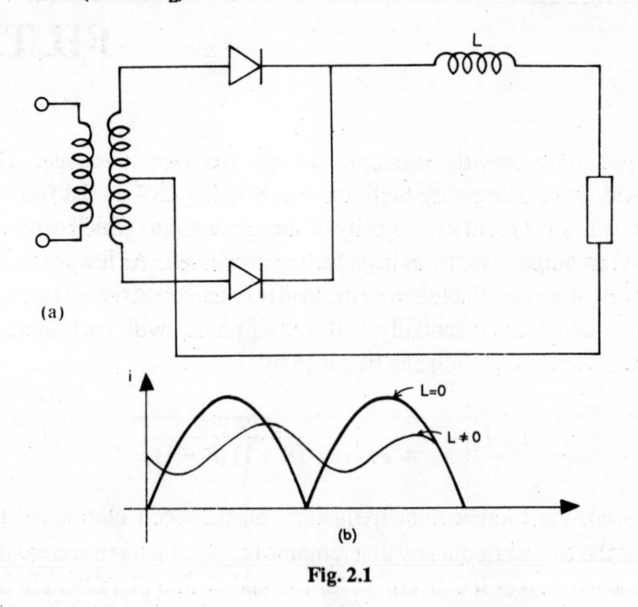

Fig. 2.1

The ripple factor, which can be defined as rms value/*dc* value =

$$E = \frac{\frac{4V_P}{3\pi\sqrt{2}} \cdot \frac{1}{\sqrt{R_L^2 + 4\omega^2 L^2}}}{2V_p/\pi R_L} = \frac{2R_L}{3\sqrt{2}} \frac{1}{\sqrt{R_L^2 + 4\omega^2 L^2}}$$

$$= \frac{2}{3\sqrt{2}} \cdot \frac{1}{\sqrt{1 + 4\omega^2 L^2/R_L^2}} \tag{2.3}$$

As can be readily seen, the output voltage will contain less ripple if

(i) R_L is less, i.e., for larger load currents, and

(ii) Inductance is large, giving larger $\omega \times L/R_L$ ratio.

In case the load resistance is infinity, i.e., the output is an open-circuit, the above equation, when rewritten, gives us the ripple factor as

$$= \frac{2}{3\sqrt{2}}$$

$$= 0.471 \tag{2.4}$$

This is slightly less than the value 0.482 as derived in Chapter 1. This reduction can be attributed to the neglect of the higher frequency terms of the Fourier series. The average, or the *dc* value of the output voltage is given by

$$V_{av} = I_{dc} \times R_L = 2\, V_p/\pi \tag{2.5}$$

However, the regulation is poorer since the output voltage linearly decreases with the increase in the load current due to the resistance of the inductor resistance, apart from, of course, the resistance of the transformer and the diodes.

$$V_{av} = (2\, V_p/\pi) - i_{dc}\,(R_d + R_t + R_c) \tag{2.6}$$

where R_d = diode resistance,
R_t = transformer resistance.
R_c = choke resistance.

2.2 Capacitor Filters

The operation of this type of filter depends on the fact that a capacitor stores energy during the conduction period, and delivers this energy to the load during the non-conduction period. In this way the time during which the current flows through the load increases, thus reducing the ripple. During the positive half-cycle, the capacitor charges to a voltage = V_p, the peak value of the transformer secondary voltage. However, during the negative half-cycle, when the diode can not conduct, the capacitor will maintain its voltage, since no path is available for its discharge. The load resistance is assumed to be equal to infinity. The output load voltage, thus, is perfect *dc* voltage, and the capacitor acts as a perfect filter. It should be noted here that the voltage now available across the diode is 2 V_p because the *ac* supply voltage becomes negative maximum and the output voltage of the previous half cycle is maintained across the capacitor. Hence, the presence of the capacitor causes the peak inverse voltage across the diode to increase from V_p, in case of the rectifier circuit without filter, to a value equal to twice the peak value of the transformer voltage for the circuit using C filters. If, now, we connect a load resistance, the diode may be assumed to work as a switch, charging the capacitor during the conduction period and preventing current flow during the negative half cycle. During the negative half cycle, however, the capacitor will be forced to discharge through the load resistance. During the positive half cycle also, the diode can conduct only when the transformer voltage is more than the capacitor voltage. Let us briefly analyse the circuit given in Fig. 2.2. As can be seen from the diagram, the diode conducts for a period which depends on the capacitor voltage. During the positive half cycle, the diode will conduct when the transformer voltage becomes more than the diode voltage. This can be called the cut-in voltage. The diode stops conducting when the transformer voltage becomes less than the diode voltage. This can be called the cut-out voltage. Since the voltage impressed on the load during positive half cycle is essentially a sinusoidal voltage-neglecting the diode voltage drop-the current through the load and the diode can be given by the equation

$$i = \frac{v}{z} = v/\sqrt{\left(R_L^2 + \frac{1}{\omega^2 c^2}\right)} \tag{2.7}$$

The value of the instantaneous current is of more importance to us. The value of the current can be given as

$$i = v/\sqrt{\left(R_L^2 + \frac{1}{\omega^2 c^2}\right)} \tag{2.8}$$

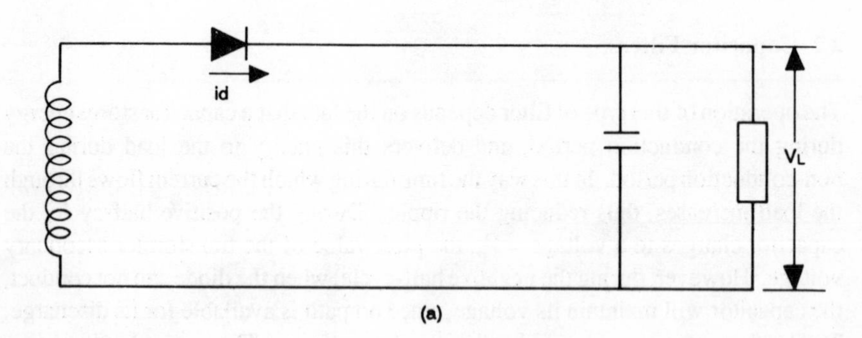

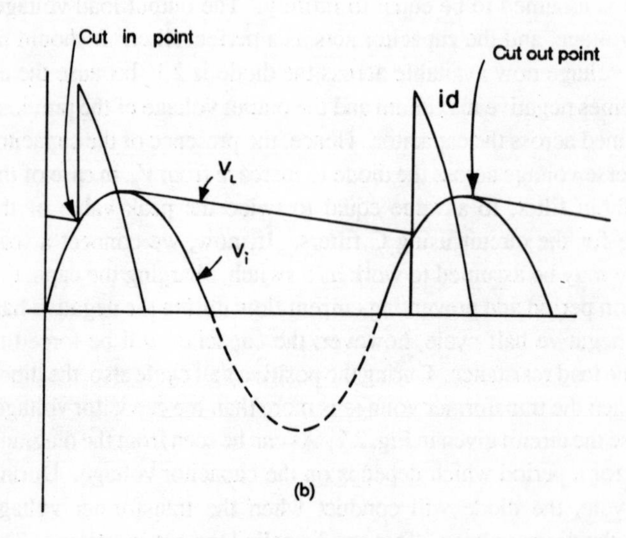

Fig. 2.2 (a and b)

It can be seen that for better filtering, the circuit will need a larger capacitor value for a given load resistance. However, a larger capacitor means a higher peak current through the diode. This current has a waveform as shown in Fig. 2.2(b). It should be emphasised that, for a given *dc* or average value of the diode current, the peak current will become larger and the conduction period will become shorter as the capacitor value is increased. This means that selecting a diode for a particular application will present greater problems. During non-conduction period, the capacitor will discharge through the load resistance, and the capacitor voltage will decrease exponentially as shown in the diagram. For a full-wave circuit, the same explanation applies, except that the discharge of the capacitor is less due to early availability of the second positive wave-form. Refer to the Fig. 2.3. Barring the first wave-form, when the capacitor voltage initially is zero, all other wave-forms are identical in shape and are as shown.

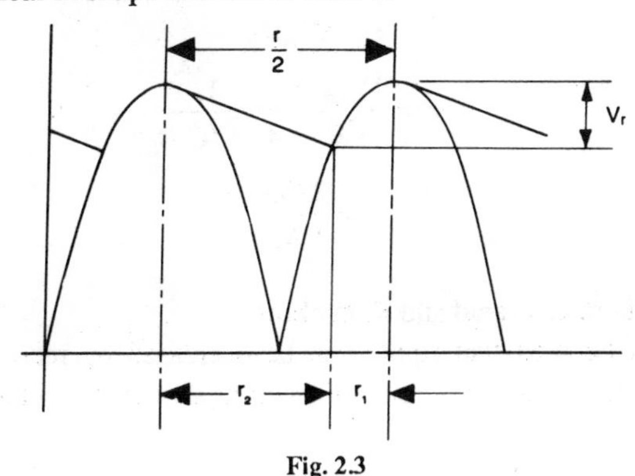

Fig. 2.3

$$V_{dc} = V_p - (V_r/2) \tag{2.9}$$

where V_r is the peak to peak ripple voltage. With slight approximation the ripple voltage wave-form can be assumed as triangular. From the cut-in point to the cut-out point, whatever charge the capacitor acquires is equal to the charge the capacitor had lost during the period of non-conduction, i.e. from cut-out point to the next cut-in point. The charge it has acquired = $V_{r\ p\text{-}p} \times C$. The charge it has lost = $I_{dc} \times T_2$

$$V_{r\ p-p} \times C = I_{dc} \times T_2 \tag{2.10}$$

If we assume that the value of the capacitor is fairly large, or that the value of the load resistance is very large, then it can safely be assumed that the conduction time is small enough to be considered negligible, and that the time, T_2 is equal to half the periodic time of the wave-form, i.e.

$$T_2 = T/2 = 1/2f, \text{ then } V_{r\ p-p} = I_{dc}/2 \cdot f \cdot C \tag{2.11}$$

With the above assumptions, and the assumption that the ripple voltage is triangular in nature, the *rms* value of the ripple will be given by

$$V_{r,rms} = \frac{V_{r,pp}}{2\sqrt{3}} \tag{2.12}$$

Therefore, from equations 2.11 and 2.12, we have

$$V_{r,rms} = \frac{I_{dc}}{4\sqrt{3}\,fC}$$

$$= \frac{V_{dc}}{4\sqrt{3}\,fC\,R_L} \tag{2.13}$$

since $$I_{dc} = V_{dc}/R_L$$

Hence, ripple $$r = \frac{V_{r,rms}}{V_{dc}} = \frac{1}{4\sqrt{3}\,fC\,R_L} \tag{2.14}$$

$$= \frac{2900}{CR_L} \tag{2.15}$$

if $f = 50$ Hz, C in micro-farad and R_L in ohms

Thus, it can be seen that ripple can be decreased by increasing either the capacitor or the load resistance, or both, resulting in increase in the *dc* value of the output voltage.

Also since $V_{dc} = V_p - \frac{V_{r,\,pp}}{2}$, from equation 2.9, we have,

$$V_{dc} = V_p - \frac{I_{dc}}{4fC} \tag{2.16}$$

from which

$$V_{dc} = \left(\frac{1}{1+\frac{1}{4fR_L C}}\right) \cdot V_p \tag{2.17}$$

It should be noted that during the conduction period, the total current through the diode is equal to the sum of currents I_c and I_L

i.e., $$I_d = I_c + I_L$$

$$= \omega \cdot C \cdot V_p \cos \omega t + \frac{V_p}{R_L} \cdot \sin \omega t \tag{2.18}$$

The equation holds good only after the conduction starts, i.e., after a cut-in point. Since the equation is frequency-dependent, we can find a maximum value of the current, i.e., the peak-current, that will be passing through the circuit. It could be a relatively simple mathematical exercise to find a maxima, which can be shown to be approximately equal to

$$i_p = i_{cp} + i_L$$

$$= \omega C V_p \frac{2.63\sqrt{r}}{1+\sqrt{3}\,r} + \frac{V_p}{R_L} \cdot \sin \omega t_1 \qquad (2.19)$$

where t_1 is the cut-in point.

2.3 LC or L-Section Filter

It was shown that the ripple increased with the increase in the load resistance in inductor filter, and decreased in case of capacitor filter. If both are combined together, then it can be reasonably deduced that the ripple can be made independent of the load resistance. Refer to Fig. 2.4. It will be interesting to study this circuit. If the inductor is not present, the capacitor will be charged to the peak value of the transformer voltage, and short high-peak, pulses of current will flow through the diodes, enough to replenish the charge in the capacitor which it loses during the non-conduction period.

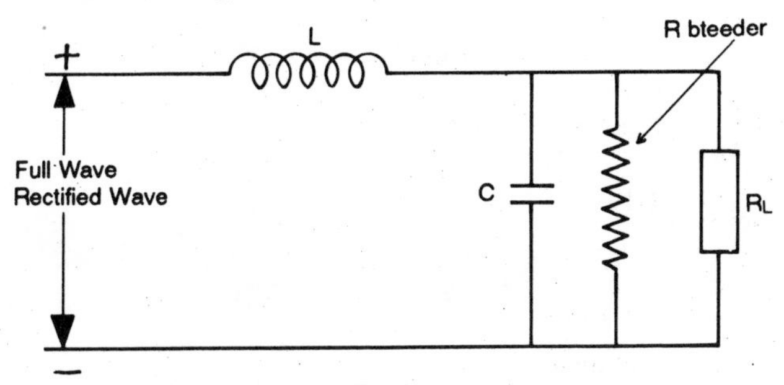

Fig. 2.4

If we introduce an inductor, it should try to smoothen these current pulses, with consequent increase in the time of conduction, but with reduced amplitude. If the inductance is increased further, it will increase the time of conduction, until at some critical value of inductance, one diode, either D_1 or diode D_2, will always be conducting. From Fourier series, we can say that the voltage will be equal to

$$v = \frac{2V_p}{\pi} - \frac{4V_p}{3\pi} \cdot \cos 2\,\omega\, t \qquad (2.20)$$

The *dc* output voltage is equal to $\frac{2V_p}{\pi}$, neglecting all resistances,

Or
$$v = \frac{2V_p}{\pi} - I_{dc}R \tag{2.21}$$

where R is the total resistance, inclusive of the resistances of the inductor, the transformer and the diode. Assuming C to be very large, and hence, X_c which is connected in parallel to the load resistance, to be very small, will determine the *ac* current essentially, by $X_L = 2\,\omega\,L$.
From the Fourier series, therefore, and equations (2.20) and (2.21), we have

$$I_{rms} = \frac{4V_p}{3\pi\sqrt{2}} \cdot \frac{1}{X_L} = \frac{\sqrt{2}\cdot V_{dc}}{3\cdot X_L} \tag{2.22}$$

This current flowing through X_c, creates the ripple voltage in the output, thus,

$$V_{r,rms} = I_{rms} \times X_c = \frac{\sqrt{2}}{3} V_{dc} \frac{X_c}{X_L} \tag{2.23}$$

The ripple factor $= V_{r,rms}/V_{dc}$

$$= \frac{\sqrt{2}}{3} \cdot \frac{X_c}{X_L} = \frac{\sqrt{2}}{3} \cdot \frac{1}{4\,\omega^2 C\,L} \tag{2.24}$$

Since $X_c = \frac{1}{2\,\omega\,C}$ and $X_L = 2\,\omega\,L$

$$\text{Ripple factor} = \frac{1.194}{LC} \tag{2.25}$$

where L is in Henry,
C is in micro-farad.
and f is 50 Hz

2.4 Critical Inductance

All the above calculations are based on the premise that the value of the inductance is such that it does not reduce the current, at any time, to zero. This would mean that the negative peak of the *ac* current must always be less than *dc*.

i.e.,
$$\sqrt{2}\cdot I_{rms} \le \frac{V_{dc}}{R_L}$$

Hence $$\frac{2V_{dc}}{3X_L} \leq \frac{V_{dc}}{R_L} \text{ from equation 2.22}$$

wherefrom $$X_L \geq \frac{2R_L}{3} \tag{2.26}$$

or $$L_{CR} = R_L/3\,\omega \tag{2.27}$$

It must be remembered that the value of the critical inductance has been based not upon the true input voltage, but rather upon an approximate voltage made up of *dc* term and the first frequency dependent term in the Fourier series of the input voltage. The neglect of the higher terms of the series introduces an appreciable error in the calculation of the critical inductance. It is, therefore, recommended that the value L_{CR} calculated should be increased by about 20 to 25%. It should also be noted that the condition that

$$X_L \geq 2\,R_L/3$$

can not be satisfied for all load requirements, since at no-load current, i.e., when the load resistance is infinity, the value of the inductance also will tend to be infinity. To overcome this problem, it is customary to introduce a 'bleeder' resistance in parallel with the load resistance. A minimum current will, therefore, be always present.

The other method would be to use a reactor which will be able to change its inductance with the variation of the *dc* current, decreasing the value of the inductance by forcing it to operate near its saturation point by the increase in the *dc* current. Figure 2.5 indicates the variation of the output voltage with the load current, which, when zero, forces the output voltage to be equal to V_p, since inductance is not effective and only capacitor filter is present.

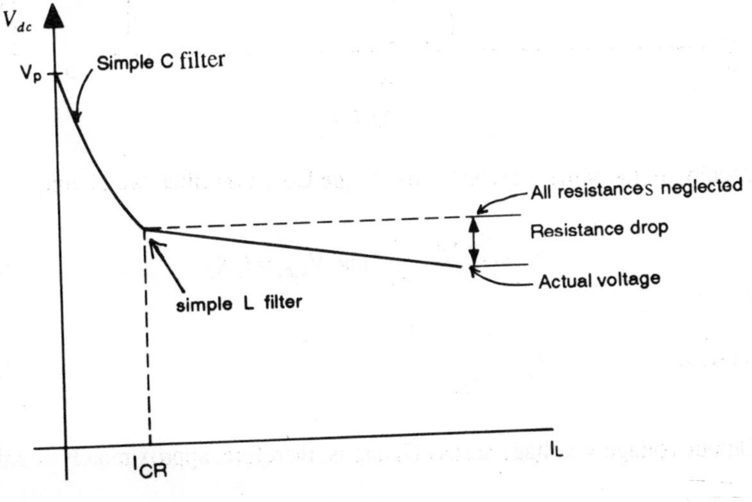

Fig. 2.5

EXAMPLE 2.1

To illustrate the above analysis, let us design a filter for full-wave rectifier circuit with LC filter to provide a 9 V current at 200 mA, ripple to be limited to 2%.

The effective load resistance $= \dfrac{9\text{ V}}{200\text{ mA}} =$ 45 ohms

since ripple factor is .02 $= \dfrac{1.194}{LC}$, giving,

$LC = 1.194/.02 = 59.7$

Again L_{CR} = Critical value of $L = \dfrac{R_L}{3\omega} = \dfrac{45}{3 \times 2\pi f} = 47.7$ mH

We may take L = 60 mH choke (about 20% higher) giving C about 1000 microfarads.

2.5 Multiple LC Filters

Better filtering can be achieved using two or more L section filters as shown in Fig. 2.6

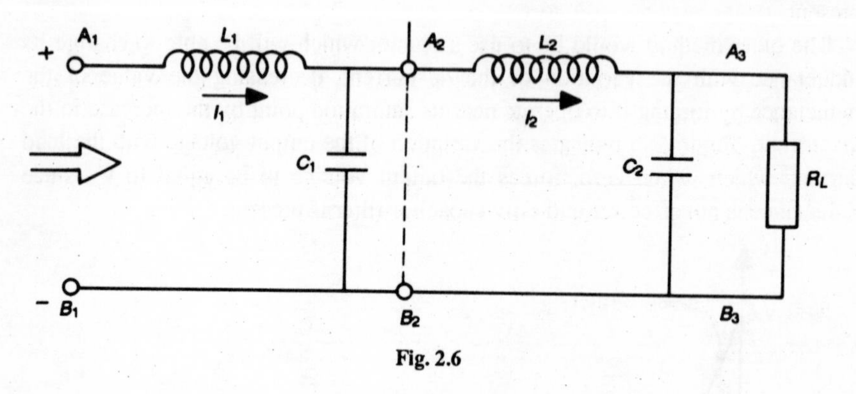

Fig. 2.6

Proceeding in the same way, as for the single L-section filter, we obtain,

$$I_1 = \frac{\sqrt{2}\,V_{dc}}{3} \cdot \frac{1}{X_{L1}} \text{ and } V_{A_2B_2} = I_1 X_{C1} \tag{2.28}$$

Likewise

$$I_2 = \frac{V_{A_2 \cdot B_2}}{X_{L_2}}. \tag{2.29}$$

∴ Output voltage = voltage across C_2 and is, therefore, approximately equal to

$$I_2 \cdot X_{C2} = I_1 \cdot \frac{X_{C2} X_{C1}}{X_{L_2}} = \frac{\sqrt{2}}{3} \cdot V_{dc} \cdot \frac{X_{C_2}}{X_{L_2}} \cdot \frac{X_{C_1}}{X_{L_1}}$$

Hence

$$r = \frac{I_2 X_{C2}}{V_{dc}} = \frac{\sqrt{2}}{3} \cdot \frac{X_{C_2}}{X_{L_2}} \cdot \frac{X_{C_1}}{X_{L_1}}. \tag{2.30}$$

EXAMPLE 2.2

With the same values as before in the example for the single section filter, if we use two sections, with the values obtained, we will have

$$r = \frac{\sqrt{2}}{3} \cdot \frac{1}{.06 \times .06} \times \frac{10^{12}}{(2\pi f)^4 \times 1000 \times 1000}$$

$$= .0285 \times \frac{\sqrt{2}}{3} = ,0134$$

against .02, in case of single section.
Or for the same ripple factor, taking $L_1 = L_2 = 0.6$ Henrys
we will have

$$C^2 = 6.72 \times 10^{-9} \text{ giving}$$

$$C = 82\,\mu F \text{ say, } 100\,\mu F.$$

2.6 The CLC or π-Section Filters

Figure 2.7 shows CLC filter. This offers a fairly smooth output, and is characterised by a highly peaked diode currents and poor regulation. Proceeding the analysis in the same way as that for the single L-section filter, we obtain,

$$r = \sqrt{2}\, \frac{X_{C1} \cdot X_{C_2}}{R_L \cdot X_{L_1}}$$

The term R_L in the above equation should be noted. For 50 Hz,

$$r = \frac{5700}{LC_1 \cdot C_2 \cdot R_L} \tag{2.31}$$

If this filter is extended to include one L-section also, consisting of L_3 and C_3, than

$$r = \sqrt{2} \cdot \frac{X_{C1} \cdot X_{C2} \cdot X_{C3}}{R_L \cdot X_{L_1} \cdot X_{L_2}} \tag{2.32}$$

The analysis can be extended to include more L-sections.

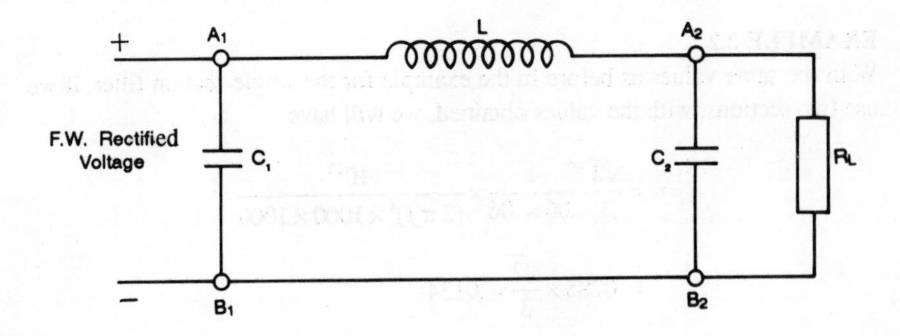

Fig. 2.7

EXAMPLE 2.3
Let us solve the example of L-section by using CLC section.

$$V_{dc} = 9V, I_L = 200 \text{ mA}, r = 2\%$$

$$R_L = 9V/200 \text{ mA} = 45\ \Omega$$

$$.02 = \frac{5700}{L \cdot C_1 \cdot C_2 \cdot 45} = \frac{126.7}{L \cdot C_1 \cdot C_2}$$

If we assume L = 10 henrys, (since L is not related to R_L as in the case of the L-section filter) and $C_1 = C_2 = C$, we have

$$.02 = \frac{12.67}{C^2}$$

giving C = 25 micro-farads.

Note: A reasonable value of L must be chosen, keeping in mind its resistance value which, if large, is going to make the regulation poorer. Refer to the example in the design section.

2.7 The R-C Filters

Figure 2.8 shows the CLC filter circuit of Fig 2.7 with L replaced by a resistance R.

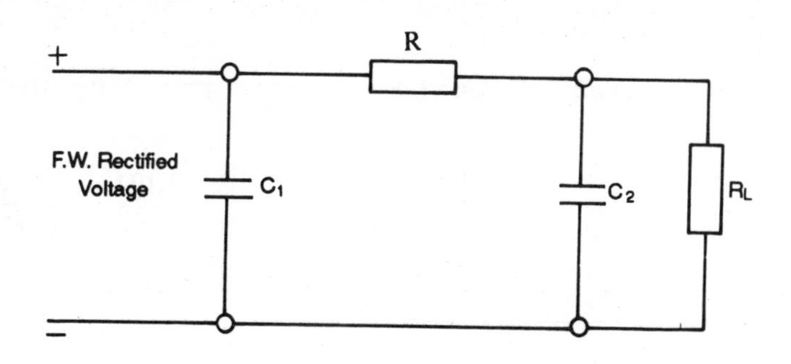

Fig. 2.8

The expression for the ripple factor can be obtained by replacing X_L by R, giving

$$r = \sqrt{2} \cdot \frac{X_{C_1} \cdot X_{C_2}}{R_L \cdot R} \tag{2.33}$$

This indicates that the values of C_1 and C_2 will be very large. The resistance R will increase the voltage drop, and hence, the regulation will be poor. This kind of filter is often used for economic reasons, as well as the space requirement of the iron-cored choke for the LC filter.

Often, this type is used to obtain a second *dc* voltage value as shown in Fig 2.9, where the load current to be supplied, usually, does not vary to a great extent.

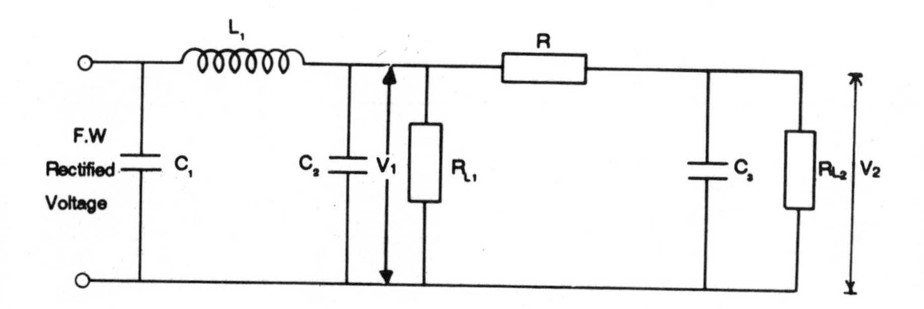

Fig. 2.9

For ready reference, all the results are tabulated in Table 2.1.

Table 2.1 Filters

	none	*L*	*C*	*L-section*	*CLC*
V_{dc} (no load)	$2V_P/\pi$	$2V_P/\pi$	V_P	V_P	V_P
V_{dc} @ I_{dc}	$2V_P/\pi$	$(2V_P/\pi - I_{dc}R)$	$(V_P - \frac{I_{dc}}{4fc})$	$(\frac{2V_P}{\pi} - I_{dc}R)$	$(V_P - \frac{I_{dc}}{4fc})$
Ripple factor, r.	0.48	$\frac{R_L}{333L}$	$\frac{2900}{CR_L}$	$\frac{1.194}{LC}$	$\frac{5700}{L \cdot C_1 \cdot C_2 \cdot R_L}$
PRV	$2V_P$	$2V_P$	$2V_P$	$2V_P$	$2V_P$

3

DESIGN OF POWER SUPPLIES

Before we attempt any design, it will be useful to know about the diode rating.

Diode Rating

The ***average current rating*** of the diode must be greater than the maximum I_{dc} expected in the diode circuit. Likewise, ***Repetitive peak current*** rating of the diode must be greater than the expected value of the I_p.

When the diode is used with shunt-capacitor filters, it is expected to supply large peaked currents. Besides, at the time of switching on the circuit, the capacitor is not charged and, therefore, will act as a short-circuit to the transformer secondary with only a diode in the circuit. The diode should be able to withstand the large inrush current. In some cases, a small resistance will have to be added to limit this surge current. In most cases, however, the transformer resistance and reactance is sufficiently large to limit the surge current. Large current diodes are manufactured by using larger junction area, which, in turn, will create larger voltage drops across the diodes, typically 1 V, and consequently power dissipation will increase. It should be emphasised here that the power dissipation during reversed biasing also is significant in such devices. Since reverse characteristic is strongly dependent on the operating temperature, the reverse power dissipation, therefore, is dependent on the operating temperatures. Large current diodes are often screwed into large surface area plates, which are used as heat-sinks, forcing better dissipation from the device to the atmosphere. However, all silicon devices have an upper limit for junction temperatures which can be as large as 175 to 200°C. But environmental temperatures play a crucial role in limiting the use of the devices. Often, 'de-rating' has to be exercised to operate the device safely.

As mentioned, semiconductor device characteristic is generally a function of temperature. As an approximation, the voltage change across a silicon diode is of the order of – 2.5 mV/°C. The change in the temperature can occur from within, due to power dissipation and/or change in the ambient temperature. Ambient temperature changes, together with the internal temperature change, will cause the voltage across the device, and hence the operating Q point, to change, or shift, depending upon the thermal resistance, θ_{JA}, between the junction-to-case and the case-to-atmosphere. Typically, these values can be assumed, respectively, between 50 to 250°C/ W, and between 5 to 25°C/ W, with a reasonably large heat-sink. As an example, a diode having $V_d = 0.7$ volts and $I_d = 1$ amp., will force a change in the voltage across the device by

$$-V = (2.5 \times 10^{-3}\ V/°C) \times 0.7\ W \times (50°C/W)$$

$$= -87.5\ \text{mV}$$

with a reasonably large heat-sink, this change reduces to

$$-8.75\ \text{mV}.$$

If the diode is kept in a small enclosure, then the surrounding area transformer may also be assumed to change, resulting in a thermal drift which, under unfavourable circumstances, may become regenerative, eventually destroying the device.

EXAMPLE 3.1

Design a full wave rectifier to supply maximum 75 ± 25 mA at 300V with a ripple less than 10V

Use: (1) L section filter (2) π-section filter

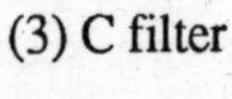

(3) C filter

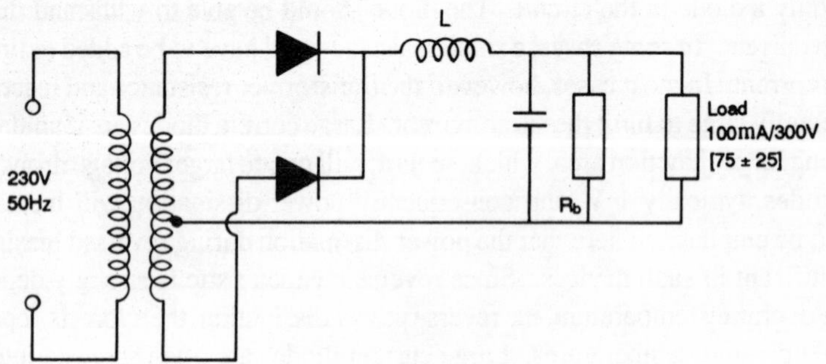

Fig. 3.1

$$I_{L(\min)} = 50\ \text{mA and}$$

$$I_{L(\max)} = 100\ \text{mA}$$

$$r = \frac{10}{300} = \frac{I_{ac}}{I_{dc}}$$

$$= \frac{V_{ac}}{V_{dc}} = \frac{1.1931D}{LC}$$

$\therefore LC = 35.8$, where L is in henrys and C in μF.

$$L_{\text{critical}} \geq \frac{R_L}{3\ \omega_s} \text{ ($\because$ min. value of $I_L = 50$ mA, we use R_B for 50 mA only)}$$

$$R_B \cong \frac{V}{I_L} = \frac{300}{50 \times 10^{-3}} = 6\ \text{K}\Omega$$

$$\therefore L_{\text{critical}} = \frac{6\,K}{3 \times 2\pi \times 50} = 6.4\ \text{H}$$

Taking 25% more, because for formula of L_{critical}, we use fundamental frequency value.

$$L_{\text{critical}} = 8\ \text{H}$$

$\therefore$ We choose $L = 10$ H/100 mA

$\therefore$ $C = 3.58$ μF from the equation.

We choose 4 μF/450 V

Now, $V_{dc} = \dfrac{2\,V_m}{\pi} - I_{dc}\,R$, where

$$\begin{aligned} R_x &= R_{\text{diodes}} + R_{\text{secondary}} + R_{\text{choke}} \\ &= 0 + 250 + 200 = 450\ \Omega \end{aligned}$$

$$V_m = \frac{\pi}{2}[V_{dc} + I_{dc}\,R]$$

$$= \frac{\pi}{2}[300 + 0.1 \times 450]$$

$$= \frac{345\,\pi}{2} = 541.92\ \text{V}$$

$$V_{\text{rms}} = \frac{542}{\sqrt{2}} = 383\ \text{V}$$

$\therefore$ Transformer used will have 390 – 0 – 390 V/100 mA.

Diode current rating = $I_L/2$ = 100/2 = 50 + 15% due to variation
= 57.5 mA.

Nearest current rating of diode = 100 mA

$\therefore$ 100 mA current rating diodes are used.

$PIV = 2V_m$ $\therefore$ PIV rating = 1084 + 15%
≈ 1400 V

Using π-section filter:

R_B is not required because there is already a path for discharge current, of capacitor, as 50 mA is minimum I_L

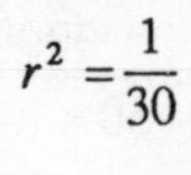

$$r^2 = \frac{1}{30}$$

also,
$$r^2 = \frac{5700}{C_1 C_2 L\, R_L} = \frac{1}{30}$$

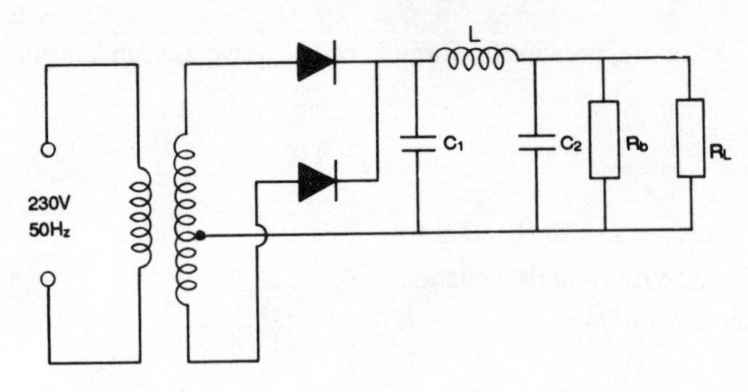

Fig. 3.2

Let $C_1 = C_2 = C$

$\therefore$
$$r^2 = \frac{5700}{C^2 L\, R_L}$$

$$R_{L(\min)} = \frac{300\text{ V}}{100\text{ mA}} = 3\text{K}\Omega,\; R_{L(\max)} = \frac{300}{50} = 6\text{K}\Omega$$

Calculating for worst case of supply, or the lowest value of $R_L = 3\text{ K}\Omega$, we have

$$\frac{5700}{C^2 L \times 3\text{ K}} = \frac{1}{30} \qquad \therefore LC^2 = 57.02$$

we select $\quad$ **$L = 1$ H/100 mA** $\qquad \therefore C^2 = 57.02$

$$\therefore C = 7.55\ \mu\text{F}$$

Using $\quad$ **$C = 8\ \mu$F/400 V.**

$$V_m = V_{dc} + I_{dc}\left\{\frac{\pi}{2\,\omega_s C} + R_L\right\} = 300 + 100\left\{\frac{\pi}{2\times 2\times \pi \times 50 \times 8 \times 10^{-6}} + 450\right\} \times 10^{-3}$$

$$= 300 + 107.5$$

$$= 407.5 \approx 408\text{ V}$$

$$V_{rms} = \frac{V_m}{\sqrt{2}} = \frac{408}{\sqrt{2}} = 288.5$$

$$= 289.$$

We select transformer as 290–0–290 V/100 mA

Voltage across C_2 = 300 V. If load goes down to 50 mA the voltage increase, is given by $V_{in} = V_{dc} + I_{dc}\left(\frac{\pi}{2\,\omega_s\,C} + R\right)$

$\therefore$ $V_{dc} = 354$, $I_{dc} = 50$

$\therefore$ voltage across C_2 in worst case is 354 V.

We select 354 + 15% = 408 V

$\therefore$ Using C_2 of 8 µF/450,

$$\text{Voltage across } C_1 = 408 + I_L\,(200\ \Omega),\ \text{where } I_L = 50\ \text{mA}$$

$$= 408 + 10 = 418\ \text{V}$$

$$V_m = \frac{\pi}{2}\{300 + 0.1 \times R_s\} = \frac{\pi}{2}\{300 + 0.1 \times 450\}$$

$$= 541.9 = 542\ \text{V}$$

$$V_m = \frac{\pi}{2}\{V_{dc} + (0.05 \times 450)\}$$

$$V_{dc} = 336\ \text{V}/50\ \text{mA} + 15\%\ \text{increase}$$

$$= 386\ \text{V}$$

$\therefore$ 450 V is needed for C

EXAMPLE 3.2

(a) Design a full wave bridge type rectifier to give O/P 15 V to a 15Ω resistance load having ripple factor better than 0.05, employing (i) L-section, (ii) π filter.

(b) Calculate the expected ripple voltage for the designed circuit. Give specifications for all components.

(c) Calculate the expected change in O/P voltage if the load resistance changes from 15Ω to 20Ω, assuming *ac* input line voltage to be constant.

(a) (i)

$$\left.\begin{aligned} \text{Ripple factor, } r &\le 0.05 \\ R_L &= 15\Omega \\ V_{dc} &= 15\ \text{V} \end{aligned}\right\}\ \text{DATA}$$

$$I_L = \frac{15}{15} = 1\ \text{A}$$

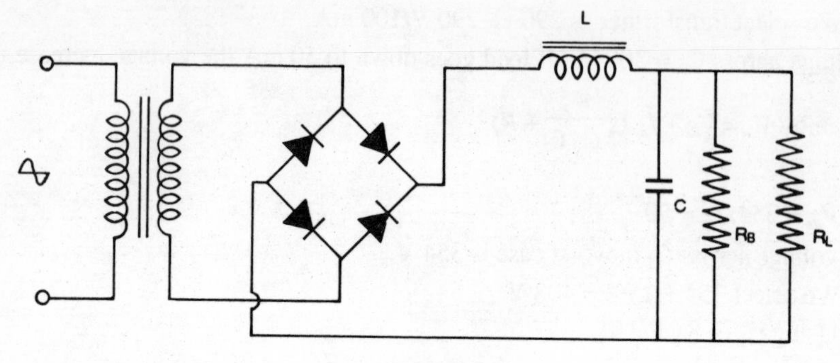

Fig. 3.3

For L-section,

$$\text{Ripple factor, } r = \frac{\sqrt{2}}{3} \cdot \frac{X_C}{X_L} = \frac{\sqrt{2}}{3} \cdot \frac{1}{\omega^2 LC}$$

$$\omega = \text{ripple frequency } = 4\pi f, \; f = 50 \text{ Hz}$$

$$\therefore \quad r = \frac{\sqrt{2}}{3} \cdot \frac{1}{(4\pi f)^2 \cdot LC}$$

$$0.05 = \frac{\sqrt{2}}{3} \times \frac{1}{(4\pi \times 50)^2} \cdot \frac{1}{LC}$$

$$\therefore \quad LC = 2.388 \times 10^5 \tag{1}$$

$$L_{\text{critical}} \geq \frac{R_L}{3\,\omega_s} \geq \frac{15}{3 \times 2 \times \pi \times 50}$$

$$\geq 15.91 \text{ mH}$$

We choose, $L = 100$ mH

$\therefore$ from equation 1, we have, $C = 238.8$ μF

We select $C = 300$ μF

(voltage rating of capacitor will be calculated later)

$\therefore$ ripple factor for $L = 100$ mH and $C = 300$ μF is

$$r = \frac{\sqrt{2}}{3} \times \frac{X_C}{X_L} = \frac{\sqrt{2}}{3} \times \frac{1}{\omega^2 L\, C}$$

$r = 0.0398$ which is better than the required value of 0.05

Inductance:
The value of L is 0.1 H (100 mH) and current rating 1 A (load current); we assume its resistance to be 1 Ω

For Transformer Selection:
Voltage drop across the choke (L)

$$= 1\ \text{A} \times 1\ \Omega = 1\ \text{V}$$

$$V_{dc}\ \text{(load)} = 15\ \text{V}$$

∴ At the input of the LC filter

$$V_{dc} = 15 + 1 = 16\ \text{V}$$

∴ Max. value of *ac* voltage at the secondary terminals required is

$$V_m = \frac{\pi}{2} V_{dc} = \frac{\pi}{2} \times 16 = 25.133\ \text{V}$$

Allowing for a 20% drop, we select

$$V_m = 1.2\ (25.133)$$

$$= 30$$

$$\therefore V_{rms} = \frac{V_m}{\sqrt{2}} = 21.2\ \text{V} \approx 22\ \text{V}$$

∴ We select a transformer of turns ratio

230 : 22

Current rating : 1.2 A
∴ V_A rating of the transformer

$$= 22(1.2)$$

$$= 26.4\ \text{VA}$$

Selection of Diodes:
PIV required $= V_m$ (max secondary transformer voltage)

$$= \sqrt{2}\ (22)$$

$$= 31.1\ \text{V} \approx 32\ \text{V}$$

Current rating = 1 A

Peak current required

$$i_{peak} = I_{dc} + \left(\frac{2}{3X_L} \cdot V_{dc} \right)$$

$$= 1 + \left[\frac{2}{3} \times \frac{15}{4 \times \pi \times 50 \times 0.1} \right]$$

$$= 1.16 \text{ A}$$

We select *IN* 4002 ≈ *PIV* = 100 V

Avg. forward current = 1 A

Repetitive peak forward current = 10 A

Selection of Bleeder Resistance (R_b):

$$L_{max} = 100 \text{ mH}$$

$$R_b \leq 3\,\omega L_{max}$$

$$\leq 3 \times 2\pi \times 50 \times 0.1$$

$$\leq 94.2\,\Omega$$

We select a std. resistance $R_b = 91\ \Omega$

$$\text{Power rating of } R_b = \frac{(V_{dc})^2}{R_b} = \frac{(15)^2}{91}$$

$$= 2.47 \text{ W}$$

$$\therefore R_b = 91\ \Omega,\ 3\text{ W}$$

(a) (ii) π-filter:

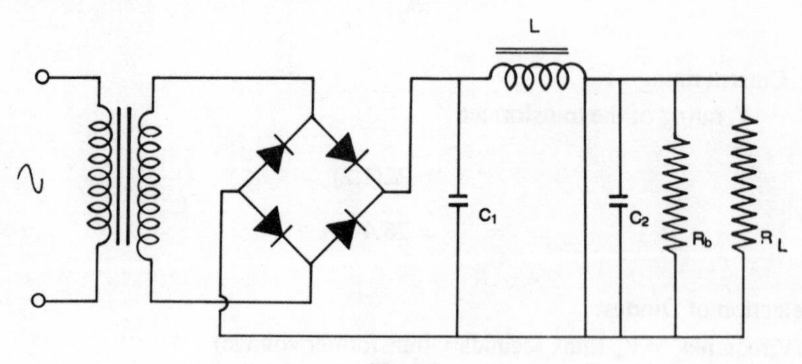

Fig. 3.4

Data

$$R_L = 15\ \Omega,\ V_{dc} = 15\text{ V},\ I_L = 1\text{ A}$$

and $$r \le .05$$

Ripple factor $$r = \frac{\sqrt{2}\, X_{C1} X_{C2}}{R_L \cdot X_L}$$

If $$C_1 = C_2, \quad f = 50 \text{ Hz}$$

and $$R_L = 15 \text{ ohms}, \quad LC^2 = 7.602 \times 10^{-9} \qquad (1)$$

Also $$L_{critical} \ge \frac{R_L}{3\,\omega}$$

$$\ge 0.0159$$

If we select $L = 100$ mH, then from equation 1, we have

$$C^2 = 7.602 \times 10^{-8}$$

giving $$C = 275.7\ \mu\text{F}$$

we select $$C = 300\ \mu\text{F}$$

Recalculating ripple factor with

$L = 0.1$ H and $C = 300\ \mu$F, we get

$$r = \sqrt{2} \cdot \frac{1}{(4\,\pi f)^3 c^2 L} \cdot \frac{1}{R_L}$$

$= 0.0422$ which is less than .05, the required value.

Assuming resistance of the choke = 1 ohm, the voltage drop across the choke

$$= 1 \text{ A} \times 1 \text{ ohm} = 1 \text{ volt.}$$

Now

$$V_{dc} = 15 \text{ V}, \text{ (at the load)}$$

$\therefore V_{dc}$ at the input of the filter section

$$= 15 \text{ V} + 1 \text{ V} = 16 \text{ volts.}$$

The peak value, V_p, at the transformer secondary can now be calculated, from

$$V_p = V_{dc} + \left(\frac{I_{dc}}{2fC}\right)/2$$

$$= 16 + \left(\frac{1}{2 \times 50 \times 300 \times 10^{-6}} \right) \times \frac{1}{2}$$

$$= 32.67 \text{ V}$$

In the absence of the actual values of the resistance of the transformer secondary, we may take these drops to be equal to about 20%, which will require the transformer secondary voltage to be equal to

$$1.2 \times V_p = (32.67) \times 1.2$$

about 39 volts.

$\therefore$ V_{rms} at the secondary of the transformer

$$= 39/\sqrt{2} = 27.6 \text{ volts.}$$

Transformer voltage ratio, therefore, should be

230 V : 28 V.

with a VA rating of the transformer

$$= 28 \text{ V} \times 1 \text{ A}$$

$$= 28 \text{ VA}$$

We should provide for slightly higher rating, say, 20% more, giving VA rating

$$= 28 \times 1.2 = 33.6 \text{ VA.}$$

Voltage rating of the capacitor will be equal to the peak value of the secondary voltage

$$= 28 \times \sqrt{2} = 40 \text{ Volts}$$

Selection of Diodes:

$$I_f = 1.2 \text{ Amps}$$

and

$$PIV = V_p = 40 \text{ V}$$

Selection of Bleeder:

$$R_b \leq 3\,\omega L$$

$$\leq 94 \text{ ohms, for } L = .1 \text{ H}$$

We select $R_b = 91$ ohms, the nearest lower standard value, having its power rating of

$$\frac{(V_{dc})^2}{R_b} = \frac{(15)^2}{91} = 2.47 \text{ W}$$

Therefore $\qquad R_B = 91$ ohms, 5 W.

Ripple Voltage:

(i) Ripple voltage (rms) $= r \times (V_{dc})$

for L section $= .0398 \times 15$

$= 0.597$ volts.

(ii) Ripple voltage for

π section $= 0.633$ volts.

It will be interesting to study the effect of change in the load current on the output voltage.

Assume that load resistance is increased from 15 ohms to 20 ohms. Then we have

(i) **L Section Filter:**

$$r = \frac{\sqrt{2}}{3} \cdot \frac{X_{C1}}{X_L}$$

Since ripple does not depend on R_L, we shall calculate the critical value of inductance

$$L_{\text{critical}} \geq \frac{R_L}{3\,\omega} \geq 21.2 \text{ mH.}$$

which is less than the selected value of 100 mH.

The output voltage remains 15 V.

(ii) **π-Filter:**

$$V_p = V_{dc}\text{ (load)} + I_{dc} \cdot R_{\text{ind}} + \left(\frac{I_{dc}}{2fC}\right)/2$$

But
$$I_{dc} = \frac{V_{dc}}{R_L}$$

$$\therefore V_p = V_{dc}\text{ (load)}\left\{1 + \left(R_{\text{ind}} + \frac{1}{4fc}\right) \cdot \frac{1}{R_L}\right\}$$

$$\therefore 32.67 = V_{dc}\text{ (load)}\left\{1 + \frac{1}{20}\left(1 + \frac{1}{4 \times 50 \times 300 \times 10^{-6}}\right)\right\}$$

$$\therefore V_{dc}\text{ (load)} = 17.35 \text{ volts.}$$

Thus, the output voltage increases with increase in load resistance. In the extreme case, when the load resistance is open-circuited,

$$V_{dc} = 32.67 \text{ volts } = V_P$$

EXAMPLE 3.3

Design a FW circuit using bridge rectifier configuration with two sections of L-C filter; to give ripple better than .03%. Inductances available are 100 mH, 100 mA, (r = 100 ohms.)

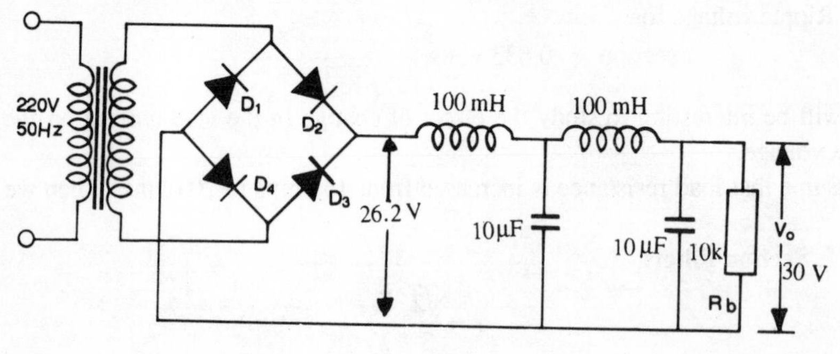

Fig. 3.5

For two L-section filter,

$$\text{ripple factor} \quad = r = \frac{\sqrt{2}}{3} \cdot \frac{1}{(4\,\omega^2 LC)^2}$$

(i) $L_{\text{critical}} \geq \dfrac{R_L}{3\,\omega}$

Since $L = 100$ mH, $f = 50$ Hz,

$R_L \leq 3\,\omega L$

≤ 9424.78

(ii) The actual value of R_L will depend on the load itself, and the output voltage required, which is not specified. We assume that $R_L = 1$ K, and the current required is 30 mA.
The load voltage = 1 K × 30 mA = 30 volts.
The supply specifications, therefore, are 30 V at 30 mA, with ripple less than .03% and operated from 220 V, 50 Hz mains.

(iii) Let us find the value of C.

Since
$$r = \frac{\sqrt{2}}{3} \cdot \frac{1}{(4\omega^2 LC)^2}$$

$$\frac{.03}{100} = \frac{\sqrt{2}}{3} \cdot \frac{1}{(4 \times \omega^2 \times .1 \times C)^2}$$

For $f = 50$ Hz, $L = .1$ H
We get, $C = 10$ uF with a voltage rating depending upon the peak-voltage of the secondary.

(iv) Transformer:
The *dc* resistance of the choke given is 100 ohms.
∴ voltage across the capacitor of the first section

$$= 30 \text{ V} + I_{dc} \cdot R_{\text{choke}}$$

$$= 30 \text{ V} \times 30 \text{ mA} \times 100 \text{ ohms}$$

$$= 33 \text{ V}$$

∴ The voltage at the input of the first section

$$= 33 \text{ V} + 40 \text{ mA} \times 100 \text{ ohms}$$

$$= 33 \text{ V} + 4 \text{ V} = 37 \text{ V}.$$

Note that the current flowing is assumed as 40 mA instead of 30 mA to take care of the capacitor current as well. (Value assumed is slightly on the higher side)
Both capacitors C_1 and C_2 can have voltage ratings of 50 V.
Transformer secondary voltage (RMS)

$$= 37/\sqrt{2} = 26.2 \text{ V}$$

Allowing for two diode voltage drops, we may require the voltage on the secondary

$$= 26.2 + 2 \times .6 = 27.4 \text{ V}.$$

This does not take into account the drop due to transformer secondary, which will depend on the rating of the transformer, but we may assume the same to be about 3 ohms for the transformer (which will create a drop of 3 ohms × 40 mA = 120 mV).
We may select transformer with turns ratio 220 : 27.5, 100 mA.

(v) The diodes should have rating of minimum 100 mA, $PIV = 60$ V.

(vi) The bleeder resistance should be around 10 times R_L, giving
$$R_B = 10 \text{ K}$$

4

REGULATED POWER SUPPLIES

In the previous chapters, we were concerned with the forward characteristics of the diodes. In fact, we ensured that the reverse voltage applied to the diode never exceeded the *PRV* of the diode, and was maintained well within the limits of the rated *PIV*. In this chapter, we will examine diodes known as the *Zener diodes*, which are operated in the reverse mode. These diodes can be used as 'stand-alone' regulator circuits, and also as reference voltage sources.

4.1 Zener Diodes

If the impurities are properly controlled, the voltage at which the break-down takes place can be varied distinctly and sharply as shown in Figure 4.1. From the characteristics, it can be seen that once the break-down occurs, practically no change in the voltage across the diode results even when the current through the diode varies. For all practical purposes, the characteristic can be assumed as a vertical line. These characteristics can be visualised as having a variable current, and an almost zero dynamic resistance. Such a diode, obviously, will have a zero resistance once it starts conduction.

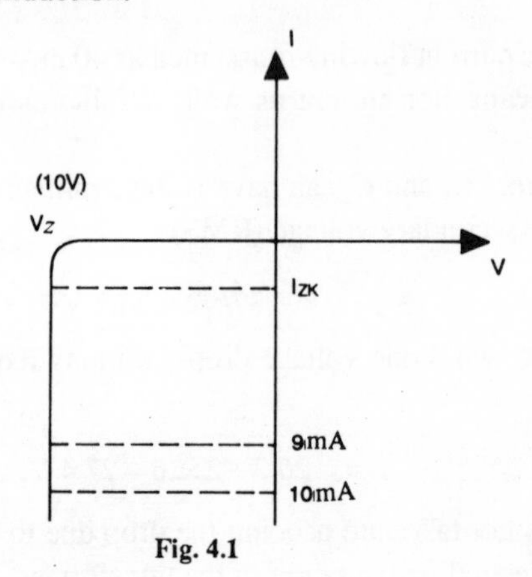

Fig. 4.1

4.2 Zener Diode as Voltage Regulator

Refer to Fig. 4.2. We have a resistance *R* connected in series with a zener diode. This series combination is connected across a voltage supply of, in our case, 20 volts. If the zener diode is assumed to start conducting at, say, 10 volts, it would mean that the voltage drop available across the series resistance is = $V_{in} - V_z = 20 - 10 = 10$ volts.

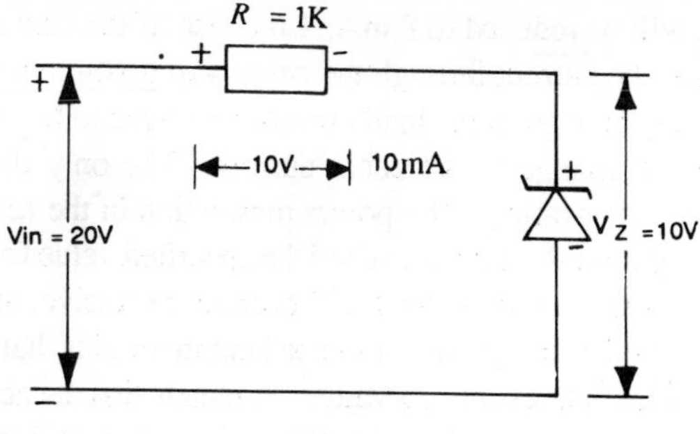

Fig. 4.2

Since we have assumed the value of the series resistance as equal to 1 K-Ohm, this would mean that a current of 10V/1k = 10 mA will flow through the zener diode.

This point of operation is shown in the diagram.

Now, if we connect a resistance R_L= 10 K-Ohms across the diode, as shown in Fig 4.3, the current through the load resistance will have to be equal to V_z/R_L = 10 V/10 K = 1 mA.

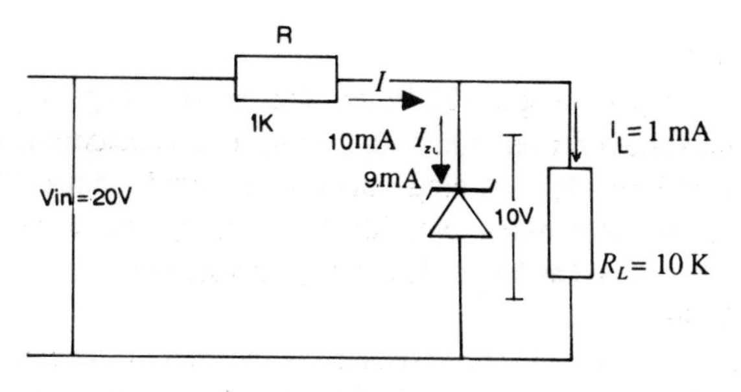

Fig. 4.3

Since the voltage across the series resistance can not change, as the input voltage and the zener voltage are assumed to be constant, this would mean that the current, *I*, flowing through the series resistance also can not change. Hence the current flow through the load resistance will have to be compensated by equivalent decrease in the zener current. Thus, the zener current will be decreased to 10 mA – 1 mA = 9 mA. We can, therefore, say that

$$I = I_z + I_L.$$

The point of operation of the zener will now change, as shown in the diagram. This change of the operating point does not mean the change in the voltage across the load resistance, which is the zener voltage, since the characteristic is almost a vertical line. If now the load resistance is decreased to, say, 5k, it would result in increase of the load current, from 1 mA to (10V / 5k) = 2 mA. Since the total current has to remain constant, it would obviously mean that the current through the zener diode will be reduced to 8 mA. Likewise, if the value of load current is increased further, the current through the zener will go on reducing. Since $I_L + I_z = I$ is constant, any change in the load current will have to be compensated by an equal and opposite change in the zener current. The only thing that has to be considered is the zener rating. The power dissipation in the zener diode is $V_z \times I_z$ Watts. This dissipation should not exceed the specified value for that diode, since, otherwise, the temperature generated will become excessive, and the device may be destroyed. The other thing that should be kept in mind is that the point at which the diode breaks-down has some curvature. It means that the zener current should not decrease to the extent that the operating point enters this region, since in this region the zener diode does not have the regulating ability. The zener voltage in this region, obviously can not have the specified rated value. Therefore, the diode is required to be operated at a current value greater that I_{zk}. This current, in most cases, is between 2mA and 5mA. Connecting the load resistance, therefore, the zener current should not get reduced below this point.

EXAMPLE 4.1

Refer to Fig 4.2.

Assume that the input voltage is 20 Volts. The output voltage required is 6 Volts at load current variation from 10mA to 20mA. Let us determine the value of R and the rating of the diode, assuming that the minimum I_z is 5mA. This means that when the load current is maximum, 20mA, the zener current will be minimum, which we have assumed as 5mA. The voltage across the resistance is $= V_{in} - V_z =$ $20 - 6 = 14$ Volts.

Since the total current in the resistance is the sum of the zener and the load currents, it will be equal to 20mA + 5 mA = 25mA. Therefore, $R = (V_{in} - V_z) / I =$ 14 / 25mA = 560 Ohms. Also the power rating of the resistance should be = 14V × 25 mA = 350 mW. We should select at least a 1/2 watt resistor. When the load current reduces to 10 mA, the zener current becomes 15 mA, as the reduction in the load current will have been absorbed by the zener. The diode capacity should therefore, be, 6V × 15mA = 90 mW. We shall be selecting the zener with much higher power dissipation capability. In case the load resistance becomes open circuited, the entire current will have to pass through the zener diode. The power dissipation of the zener in that case will be = 6V × 25mA = 150 mW. While selecting the zener, the above possibility should be kept in mind.

The load resistance in the above case changes from 6V / 10 mA = 600 Ohms to 6V / 20 mA = 300 Ohms.

In the above example we have assumed that the input voltage remains constant. However, if the input voltage changes, e.g. it increases to 25 Volts, then, since the voltage across the series resistance is $= V_{in} - V_z$, it would mean that the voltage across the series resistance increases to $25 - 6 = 19$ Volts. The current through the series resistance will increase to 19V / 560 Ohms = 33.93mA. This current was 25mA when the input voltage was 20V. Again, since the load current does not change, as zener voltage is constant, I_z becomes equal to 33.93mA – 10mA (minimum load current) = 23.93mA, which increases the power dissipation to 23.93mA × 6V = 143.58mW. This would necessitate using a larger capacity zener diode.

We may, therefore, take

$$I = I_{L(max)} + I_{z(min)}$$

and

$$I \times R = V_{in(min)} - V_z$$

It should be remembered that the V_z should be the voltage at the point of operation.

From the above equations, we can deduce that

$$R = \frac{V_{in(min)} - V_z}{I_{L(max)} + I_{z(min)}} \tag{4.1}$$

If v_{in} *is the change in the input voltage* V_{in}, and if
v_o *is the change in the output voltage* V_o then,

$$v_o = \frac{R_z}{R + R_z} \times v_{in} \text{ (refer to Fig. 4.4)}$$

or,

$$\frac{v_o}{v_{in}} = \frac{R_z}{R + R_z} = S \tag{4.2}$$

The factor, S, can be termed as the voltage stabilisation of the regulator and, obviously, should be as small as possible; ideally S should be 0, which would mean that even if the input voltage changes, the output voltage remains constant. (lower-case letters v_{in} and v_o denote the ***change*** in the values, respectively, in the *dc* values V_{in} and V_O, which are denoted by the upper-case letters.

In the above relation, if the load resistance also is considered, then, R_z should be replaced by the parallel combination of R_z and R_L. Since R_z is relatively very small as compared to R_L, the result remains virtually the same. Again referring to Fig 4.3, and Thevenizing the circuit as shown in Fig 4.4, we get the 'internal' resistance of the supply as

= R_o = parallel combination of R and R_z.

Maximum current flowing through the resistance R

$$= I_{max} = \frac{V_{in(max)} - V_{z(min)}}{R} \tag{4.3}$$

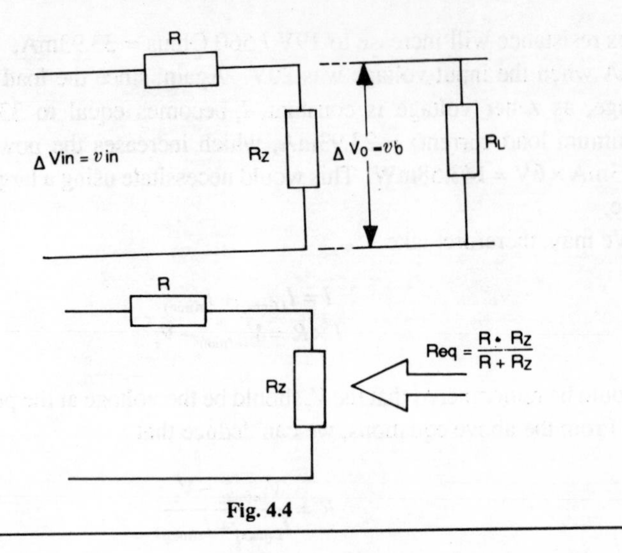

Fig. 4.4

EXAMPLE 4.2

Design a zener voltage regulator to supply a load current which varies between 10mA and 25mA at 10V. Input supply voltage available is 20V ± 20%.

Solution

Refer to Fig 4.5

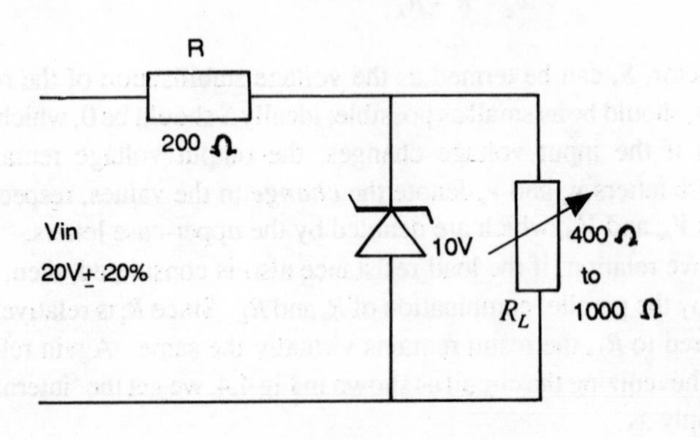

Fig. 4.5

1. Since the output voltage = 10V., the load resistance, R_L, varies from

$$V_z / 10\text{mA} = 1000 \text{ Ohms to}$$
$$V_z / 25\text{mA} = 400 \text{ Ohms.}$$

2. Given that $V_{in(max)} = 20 + 0.02 \times 20 = 24\text{V}.$

$$V_{in(min)} = 20 - 0.02 \times 20 = 16\text{V}.$$

Hence, $$R = \frac{V_{in(min)} - V_z}{I_{L(max)} + I_{z(min)}}$$

$$= (16 - 10)\text{V}/(25 - 5)\text{mA}$$

$$= 200 \text{ Ohms},$$

assuming $I_{z(min)}$ to be slightly greater than I_{zk}.

3. The zener is called upon to absorb maximum current when V_{in} is maximum and at the same time I_L is minimum.

$$I_{z(max)} + I_{L(min)} = (V_{in(max)} - V_z)/R.$$

Hence, $$I_{z(max)} = \{(24 - 10)/200\} - 10\text{mA}$$

$$= 60\text{mA}.$$

The zener diode rating, therefore, should be

$$P_{z(max)} = V_z \times I_{z(max)} = 10\text{V} \times 60\text{mA}$$

$$= 600\text{mW}.$$

We will have to select a zener diode of rating of 1W. This higher rating will also take care of the inadvertant opening of the load resistance circuit, in which case, the zener current will be equal to 70mA and the zener power dissipation will be equal to

$$P_z = 10\text{V} \times 70\text{mA} = 700\text{mW}.$$

In the above analysis, the dynamic resistance of the zener diode is assumed to be equal to zero (and, therefore, $S = 0$, as per equation 4.1). However, even if we assume $R_z = 5$Ohms (say), then V_z will change with the change in the zener current. The zener current changes from its minimum value of 5mA to a maximum of 60mA. Therefore, v_o = the change in the output voltage = (60 – 5)mA × 5 Ohms = 0.275V.

Consequently, the stabilisation factor will be =

$$S = R_z / (R_z + R) = 5 / 205 = 0.024.$$

4.3 Emitter-Follower Type Regulator

In example 4.2, we have seen that the zener current varies from **5mA** to **60mA** as the V_{in} and I_L vary.

As a result the output voltage, which is equal to V_z, also changes by 275mV.

This change in the output voltage can be minimised if we can reduce, somehow, the change in the zener current as a result of change in the load current and the change in the input voltage. One such method is depicted in Fig 4.6.

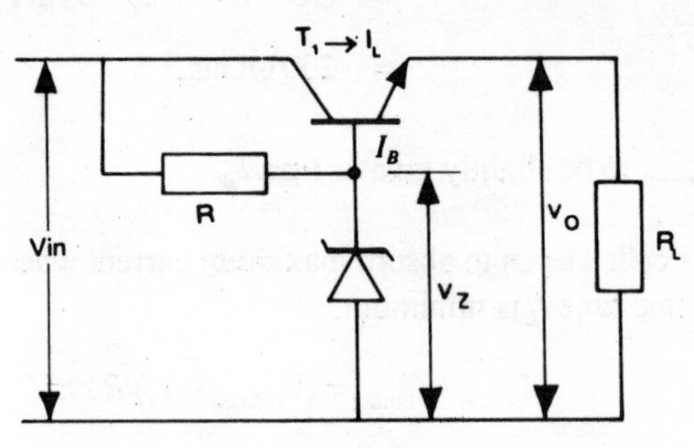

Fig. 4.6

In this method, the load resistance, R_L, is not connected across the zener directly as in the previous case, but is connected through an amplifier/buffer circuit. Transistor T_1 is connected as an emitter-follower. As can be seen, V_o voltage will be about 0.6 volts less than the zener voltage, i.e., by voltage equal to the V_{BE} drop. If the zener is of 10V, then the output voltage available will be about 9.4V.

However, the load current is supplied by the transistor T_1 from V_{in} voltage, deriving its base current from the zener circuit. The base current will be equal to I_L/β, where β is the current-gain of the transistor in question.

As before,

$$I = I_z + I_b$$

where I_b will be able to support upto ($\beta \cdot I_b$) current through the load. As far as the zener circuit is concerned, it is supplying only the base current as the load current. Any change in the load current is reduced by β times before affecting the zener current. Therefore, a change in 100mA in the load-current, for instance, will produce a change of 1mA in I_b, and, hence the zener current I_z, assuming β to be equal to 100.

EXAMPLE 4.3

Design a power supply to be capable of delivering upto 250mA at 10V.

Solution

Since the load current is fairly large, we shall employ a series-pass or emitter-follower type of circuit.

If we assume the β of the transistor = 100,

for a load current of 250mA, the base current will be

$$= I_B = 250\text{mA}/100 = 2.5\text{mA}.$$

Let V_{in} voltage be equal to about twice the output voltage,

i.e., $$V_{in} = 20 \text{ volts} \pm 20\%$$

or V_{in} varies between 16 and 24 volts.

We will also assume that the minimum zener current is = 5mA. This gives us

$$I = I_{z(\text{min})} + I_{L(\text{max})}/\beta$$

$$I = 5\text{ mA} + 2.5\text{ mA} = 7.5\text{ mA}.$$

and $$R = \frac{V_{in(\text{min})} - V_z}{I} = \frac{16-10}{7.5\text{mA}}$$

$$= 6/7.5\text{mA} = 800\text{ Ohms}.$$

However,

$$I_{z(\text{max})} + I_{b(\text{min})} = \frac{V_{in(\text{max})} - V_z}{R}$$

$$= (24-10)/800$$

$$= 17.5\text{mA}.$$

$$I_{z(\text{max})} = 17.5\text{mA} - I_{b(\text{min})} = 17.5 - 2.5$$

$$= 15\text{mA}.$$

This shows that for the full change in the load current, i.e., a change of 250mA in the load circuit, the zener current will vary to the extent of 2.5mA, from 15mA to 17.5mA. This change in the zener current will produce a change in V_z, at the most, of 2.5mA × 5 Ohms (typical R_z) = 12.5mV.

The important thing, however, is the power dissipation of the zener which is going to be = 17.5mA × 10V = 175mW.

In the above circuit, V_{BE} drop has been neglected. If it is assumed to be equal to about 0.6V, then the output voltage will be

$$V_o = V_z - V_{BE} = 10 - 0.6$$

$$= 9.4\text{V}.$$

If the output voltage is desired to be about 10V, then V_z will have to be 10.6V, which is not a standard zener diode value. This difficulty can be overcome by inserting a normal diode in the circuit as shown in Fig 4.7.

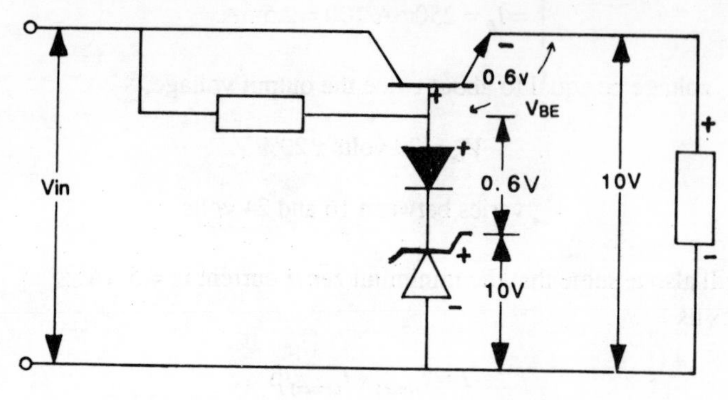

Fig. 4.7

This diode will boost up the base voltage by about 0.6V, forcing output voltage to be equal to 10V.

This will have yet another useful and desirable effect—that of temperature compensation. A zener diode having break-down voltage greater than 6V, normally has a positive temperature co-efficient i.e., its break-down voltage increases with increase in temperature. Together with the negative temperature coefficient of the normal diode, the temperature effect is, to a great extent, nullified.

4.4 Darlington Pair as Series Pass

In example 4.3, we assumed a change in the load current of about 250 mA, forcing a change in the base current of the transistor T_1 to about 2.5mA. This was with the assumption that the transistor T_1 has a current gain of about 100. The transistor, in the example, is supposed to carry 250mA with a voltage across it of about 24 – 10V, when the input voltage is maximum. This means that the power dissipation of the transistor is = 250mA × 14V = 3.5W. The transistor will have to be a medium power transistor which normally has the current-gain in the region of about 50. (Like SL100 or ECN100). Since the gain is comparatively less, it would mean larger changes in the base current for the specified load current changes. To obviate such problems, we can and should, use a Darlington pair of transistors.

EXAMPLE 4.4

Design a power supply to give 1A at 12 volts.

Solution

We shall assume an input voltage supply to be available having twice the output voltage, i.e., say, 25V, with a variation of ± 20%.

The input voltage available, therefore, will be between 20 and 30 volts.

The circuit, rearranged, will be as shown the Fig 4.8.

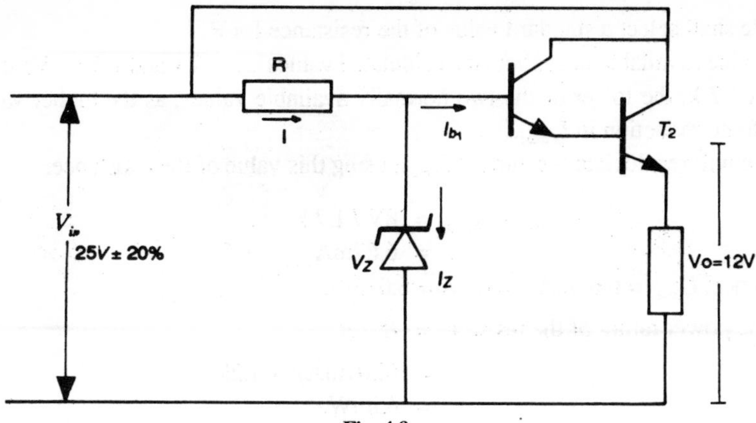

Fig. 4.8

1. Let us select, first, the transistor T_2.
   V_{CE} across the transistor $= V_{in(max)} - V_o$

   $$= 30 - 12 = 18\text{V}.$$

   Hence, the power dissipation for the transistor $T_2 = 18\text{V} \times I_L$

   $$18\text{V} \times 1\text{A} = 18\text{W}.$$

   We will select ECN149 as our transistor T_2. This transistor has a gain of about 35.

2. The base current of the transistor T_2 for full variation of the load current, is $= 1\text{A}/35 =$ about 30 mA.
   The transistor T_1 will be supplying the base current of transistor T_2. With the voltage across the transistor T_1 being practically the same as that of the transistor T_2, the power dissipation of the transistor T_1 will be, neglecting V_{BE2},

   $$= (30 - 12)\text{V} \times 30\text{mA} = 18 \times 30\text{mA} = 540\text{mW}.$$

We will, therefore, have to select a transistor which has this capacity, and also can withstand 30mA. We select transistor SL100 for the purpose, which has a current gain of about 50.

This gives $I_{B1} = 30\text{mA} / 50 = 0.6\text{mA}$.

3. This I_{B1} is the load-current for the zener diode, which is operated slightly above I_{zk} of, say, 5mA.

$$R = \frac{V_{in(min)} - V_z}{I_{z(min)} + I_{b(max)}} = \frac{20 - 12}{5.6\text{mA}}$$

$$= 8\text{V}/5.6\text{mA} = 1.42 \text{ K-Ohms}.$$

We shall select a standard value of the resistance for R.

The values available near about the calculated values are, 1.2 k and 1.5 k. We shall select 1.2 k, the lower of the two normally available values, as the higher value will force reduction in $I_{z(min)}$.

We will recalculate the current $I_{z(min)}$, using this value of the resistance.

$$I_{z(min)} + I_{B(max)} = 8V / 1.2\,k = 6.67mA.$$

Hence, $I_{z(min)} = 6.67mA - 0.6mA = 6.07mA.$

The power rating of the resistor $= I_z^2 \times R$

$$= (6.07mA)^2 \times 1.2k = 45mW.$$

We shall select a resistance of 1/4 W capacity.

4. To obtain 12 volts at the output, the zener diode will have to be of 12V + $V_{BE1} + V_{BE2} = 12 + 0.6 + 0.6 = 13.2$ V.

We can adopt the strategy shown in Fig. 4.9

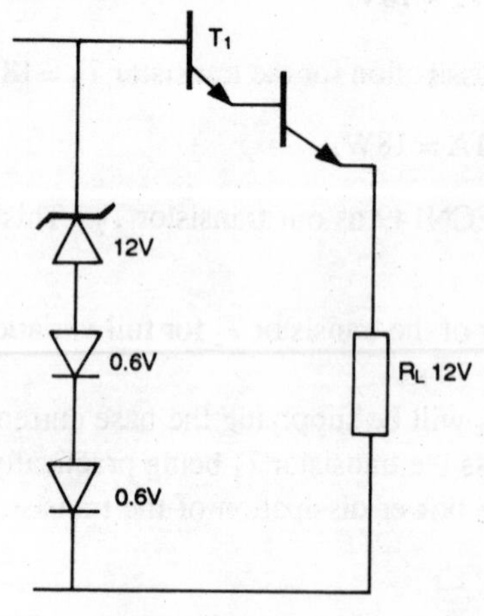

Fig. 4.9

5. The power rating of the zener diode is

$$V_z = 12V \text{ and } I_z = 6.07mA$$

Hence, $P_z = 12V \times 6.07mA = 80mW$. We can select 150mW zener. In case there is an open-circuit on the load side, the zener current will increase to 6.07mA + 0.6 mA = 6.67mA, needing the zener diode to dissipate power = 12V × 6.67mA = 80mW.

The zener diode selected can take care of the increased dissipation.

We need the following components to build the circuit.

Transistor T_1 = SL100 or ECN100.
Transistor T_2 = ECN149.

Zener diode = 12V, 150mW.
Two diodes = IN4001.

4.5 Over-load Protection

Often it is desirable to provide protection against the over-load, since active devices like transistors, are very sensitive to the over-current, and to the resulting greater power dissipation.

Refer to Fig 4.10

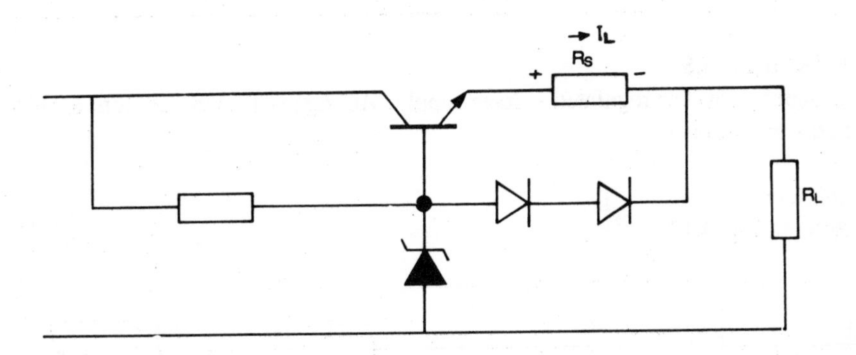

Fig. 4.10

A small resistance R_s is added in series with the load resistance. The emitter voltage is = $(V_z - 0.6)$. The load current passing through this resistance creates voltage drop across it, which is = $I_L \times R_s$.

We have connected two diodes from the base of the transistor to the output. So long as the voltage drop available across the diodes is less than twice the cut-in voltage of the diode, i.e., approximately 1.0 volt, the diodes are effectively as good as not connected in the circuit. However, as soon as the voltage drop across the resistance R_s becomes more than about IV, the diodes will be forward-biased and will start conducting. This will divert part of the base current, which will be directly led to the output, thus, restricting the base current and, hence, the transistor current. With a proper design, the transistor can be turned-off, which will be necessary in case the load-resistance is reduced drastically, as in the case of a short-circuit.

The protective diodes can be replaced by yet another transistor T_3 as shown in Fig 4.11. In this case, the drop across the resistance R_s is used in turning ON the transistor T_3, giving the same effect as before. The voltage drop across the sensing resistance will have to become greater than the cut-in voltage of the transistor. For a current limit of 1A, for example, we will need to connect a series resistance R_s = (cut in voltage of transistor T_3) / 1A

= about 0.5V / 1A = 0.5 Ohm.

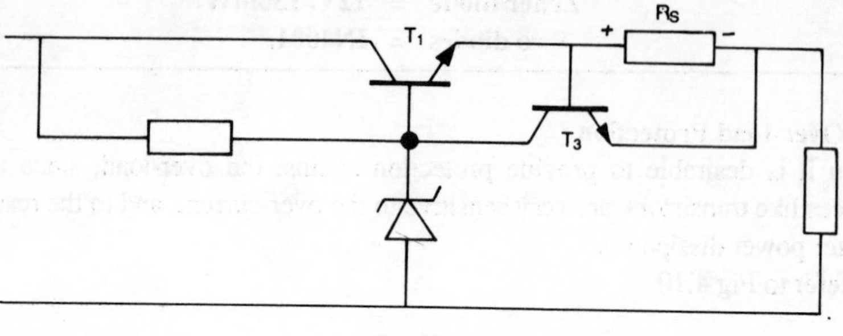

Fig. 4.11

EXAMPLE 4.5

Design 5V, 500ma regulated voltage supply. *AC* supply is obtained from a 230V : 20V transformer.

Solution

Refer to Fig. 4.12.

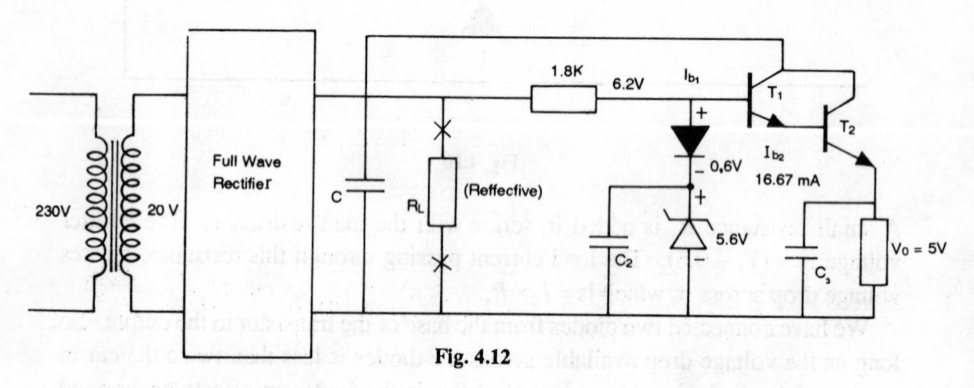

Fig. 4.12

1. The transformer secondary will have a voltage = 20V rms

$$= 20 \times \sqrt{2}$$

$$= 28.28\text{V peak.}$$

2. We will select a capacitor filter even though the load current is large.

$$V_{av} = V_p = 28.28\text{V}$$

Actually this voltage should be equal to

$$V_{dc} = \frac{(4fR_LC)}{(4fR_LC + 1)} \times V_p$$

$$= \frac{(4 \times 50 \times R_L \times C)}{(4 \times 50 \times R_L \times C + 1)} \times 28.28 \qquad \text{(i)}$$

Since this supply will have to feed, $I_L + I_r$
$= 1\text{A}$, neglecting I_r,
giving $I_{dc} = 0.5\text{A}$, and hence $R_L = V_{dc} / I_{dc}$
$= 5\text{V} / 0.5\text{A} = 10 - \text{Ohms}$
Substituting this value in the equation (i) we have,

$$V_{dc} = \frac{(2000 \times C)}{(2000 \times C + 1)} \times 28.28\text{V} \qquad \text{(ii)}$$

Again assuming a ripple in the rectified output voltage to be 5%, (this value of ripple is large, but since we are going to use a regulated supply this ripple should not matter much.)
$0.05 = 2900/C \times R_L$ (Refer to Table 2.1)

therefore, $$C = \frac{2900}{10 \times 0.05} = 5800 \text{ uF}$$

We will select two capacitors of 2500 uF, or a single capacitor of 5000 uF with voltage rating of greater than 30V, say, 50V, the standard voltage rating available.

Thus, from (ii),

$$V_{dc} = \frac{(2000 \times 5000 \times 10^{-6})}{(2000 \times 5000 \times 10^{-6} + 1)} \times 28.28\text{V}$$

$$= \frac{10}{11} \times 28.28 = 25.78\text{V}.$$

3. *Selection of the transistors.*
Let us select the transistor T_2.
V_{ce} for the transistor $T_2 = V_{in} - V_o$
$= 25.78 - 5 = 20.78\text{V}$
Since the load current will have to flow through the transistor T_2, the power dissipation capacity of transistor T_2 is

$$= V_{ce} \times I_L$$

$$= 20.78\text{V} \times 500\text{mA}$$

$$= 10.39\text{W}.$$

We select a transistor ECN149 having $I_{c(max)} = 4\text{A}$., and maximum power dissipation of 30W. This transistor has minimum h_{FE} of 30.
This will mean that a base current of 500mA / 30 = 16.67mA will be required to support the required load current through the transistor.

Selection of the transistor T_1: The above base current will be supplied by the transistor T_1. This is the collector current of the transistor T_1. Voltage across this transistor is practically the same as that of the transistor T_2. Actually it is

$$= V_{in} - V_o - V_{BE2}$$

$$= 25.78 - 5.0 - 0.6$$

$$= 19.58\text{V}$$

Since the current flowing through the transistor is = 16.67mA, the power dissipation for this transistor

$$= 19.58 \times 16.67\text{mA}$$

$$= 326\text{mW}.$$

This will necessitate the use of a transistor like SL100 as transistor T_1. This transistor has $h_{FE(\min)} = 50$.

4. *Selection of zener diode*

Transistor T_2 base is at a voltage $= V_o + V_{be}$

$= 5 + 0.6 = 5.6\text{V}.$

Likewise, T_1 base should have a voltage of $V_{b_2} + 0.6 = 6.2\text{V}$.

We can take a zener of 6.2V, or a zener of 5.6V in series with a diode to provide about 6.2V at the base of T_1.

Resistance $R = (V_{dc} - V_{b1})/I$

$$= \frac{25.78 - 6.2}{I} = \frac{19.58}{I}$$

where current $I = I_{b_1} + I_z$

$$I_{b_1} = I_L / (h_{FE1} \times h_{FE2})$$

$$= 500\text{mA} / (50 \times 30) = 0.33 \text{ mA}.$$

Hence, $R = 19.58\text{V} / 0.33\text{mA} = 1.958$ K-Ohms.

We can select a nominal value of 1.8k-ohms.

Recalculating the value of the current passing through it,

We have $I = I_z$(approx.) = (25.78 – 6.2)V / 1.8k = 10.88mA.
The value of the resistance R is thus 1.8k, and since the current passing through it = 10.88mA, the power rating of R

= $(10.88)^2 \times$ 1.8k-Ohms = 213mW.

We can select R as 1.8k-Ohms/ 1/2 W rating.

NOTE

It should be noted that the input peak voltage is assumed as 28.28 volts for a 20V rms value of *ac* supply voltage, and that, this supply voltage is assumed to remain constant. If this voltage changes by about 20%, then, the maximum voltage across the filter capacitor, i.e., at the input of the regulator will be equal to 34 volts.

Giving,

$$V_{dc} = \frac{2000\,C}{2000\,C+1} \times V_p$$

$$= (10/11) \times 34$$

$$= 30.9\text{V},$$

for a capacitor of 5000 uF, and a peak voltage of 34 volts.

For the value of the resistance R = 1.8k, the current $I = I_z$ (approximately) may increase to a value

$$= I = \frac{V_{dc} - V_{b_1}}{R} = \frac{30.9 - 6.2}{1.8\text{k}}$$

$$= 13.8\text{mA}.$$

This in itself does not present a problem, since all the components selected so far will still perform well within their limits. However, the change in the zener current from 10.88mA to 13.8mA, resulting from increase in the voltage, will produce a change in zener voltage and, hence, a change in the output voltage, by about (13.8 – 10.88)mA × 5 ohms (approximate value of the zener dynamic resistance) = about 15mV.

This entire change will appear across the load resistance in the form of a ripple voltage. The ripple voltage may also come from other sources; for instance, from reverse biased collector-to-base junction of transistor T_2 (and may be of transistor T_1 also). This is attributed to a relatively low resistance of the collector-to-base junction of the power transistors. (This resistance could be as low as, between, 8k and 15k.) This low resistance will permit part of the input voltage change to appear at the base, and, hence, at the output.

Yet another source of the ripple is the behaviour of the load itself. If the load current switches from one value to another fairly regularly, a ripple of this frequency will appear in the output voltage. This condition is very severe if the load driven happens to be a digital circuit. The most of the digital families may offer momentarily a virtual short-circuit at a moment when they change from one logic level to another. Also, a typical gate (e.g TTL NAND) draws an increased current from the supply, from 4.0mA to 12mA, when its output changes from a logic high to logic low. A system having a circuit complexity of about 100 gates will force the current from the supply to change by, on average, about 400mA !
Since this change occurs extremely fast - in nano-seconds - the regulator circuit has no time to respond to this change and, therefore, can not adjust its operation to maintain the output voltage constant. There will, therefore, be almost invariably, a spike in the supply voltage. This may cause, at worst, a malfunction in the digital system in its logic operation.

This problem can be solved by
(i) employing a capacitor C_2 of a value of between 10 and 50 uF across the zener diode to suppress the ripple caused by the changes in the input voltage, and
(ii) using a capacitor C_3 of a value of about 50 uF (for low current circuits) to 500uF (for high load-current circuits).

If the load is digital integrated circuits, then, a small value capacitor of about 0.01 to 0.1uF is connected between supply and the ground pins of each IC to by-pass the spikes, effectively 'de-coupling' the circuit.

5

SERIES VOLTAGE REGULATOR

The voltage regulation available in this type of circuit is far superior to the circuits we have discussed so far. Besides, the out-put voltage, which is fixed and is equal to V_z in the circuits discussed so far, can also be varied over a range.

5.1 Series Voltage Regulator

The circuit can be best understood as follows:
Refer to Fig. 5.1.

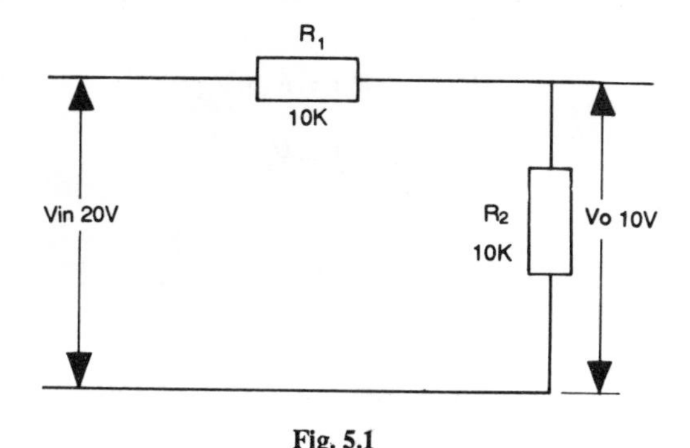

Fig. 5.1

The output voltage available across the resistance R_2

$$V_o = \frac{(R_2)}{(R_1+R_2)} \cdot V_{in} \tag{5.1}$$

$$= \frac{10}{10+10} \times 20 = 10 \text{ V}$$

If we now assume that the input voltage changes by, say, 5V to 25V, then the output voltage will also increase by

$$\frac{5 \times R_2}{R_1+R_2} = 2.5 \text{ V}$$

i.e., the output voltage will become 12.5V. However, assuming that we have some control circuit which can sense change in the output-voltage and automatically changes the value of the resistance R_1, as shown in Fig 5.2, to a value such that output voltage remains 10V as before.

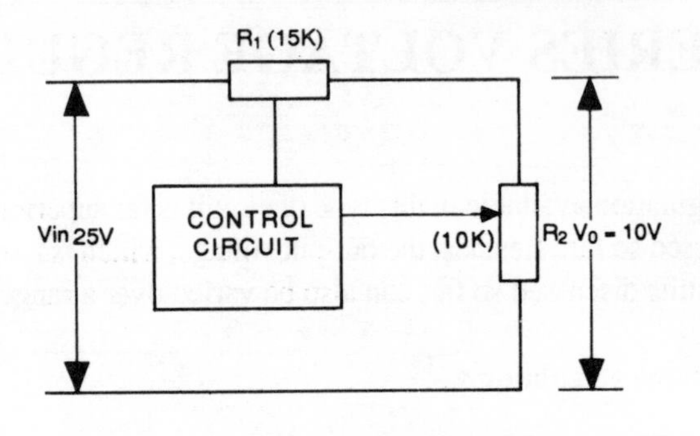

Fig. 5.2

From equation 5.1, we have

$$V_o\,(R_1+R_2) = R_2 \times V_{in}$$

or,

$$V_o\,(1+R_1/R_2) = V_{in}$$

or,

$$\frac{R_1}{R_2} = \frac{V_{in}}{V_o} - 1$$

i.e.,

$$R_1 = R_2\,(V_{in} - V_o)/V_o \tag{5.2}$$

Therefore,

$$R_1 = 10\text{K}\,(25-10)/10 = 15\text{K}$$

Hence, the control circuit should adjust the value of R_1 to 15K to keep the output voltage = 10V

It should be noted here that since the control circuit changes the resistance R_1 ***after sensing the change in the output voltage***, and since the new value of R_1 should depend on the new changed output voltage, it would mean, therefore, that when the output voltage is brought back to its original value, the resistance R_1, should also be brought back to its original value, of 10K in our case. For the control circuit to be effective, therefore, the output voltage should *remain changed, albeit*, to a small extent, and the control circuit will have to be highly sensitive, and also should have high gain. The resistance R_1 will have to be an active resistance.

Refer to Fig. 5.3.

If we consider the transistor as a resistor, then, the value of resistance offered by the transistor = V_{CE}/I_C.

If we now increase the base current I_B, then current I_C will increase, with a consequent fall in V_{CE}.

Hence the resistance = V_{CE}/I_C will decrease. Thus, by varying the current I_B, we are able to change the value of the effective resistance of the transistor, (assuming the transistor is working in the active region).

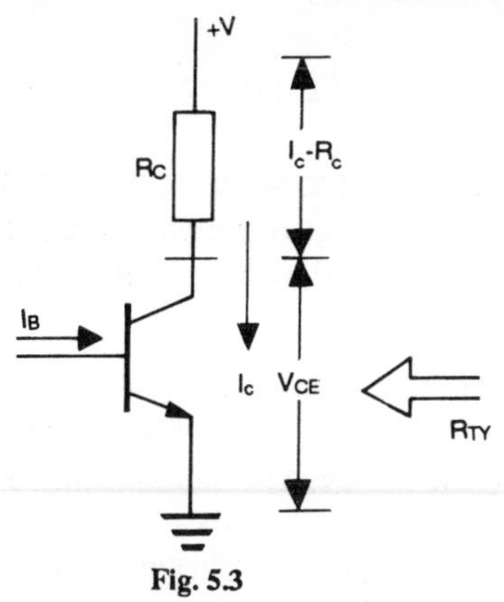

Fig. 5.3

Our control circuit, thus, has to sense the change in V_o, and correspondingly increase, or decrease, the base current of transistor T_1 which will replace the resistor R_1, as shown in Fig 5.3. The control circuit senses the change in V_o by simply comparing it with a reference voltage source, normally a zener diode, and amplifying the difference between the two.

5.2 Analysis of Series Voltage Regulator

Figure 5.4 shows a typical series voltage regulator.
Transistor T_1 replaces R_1 of the circuit shown in Fig. 5.2 and T_2 replaces the control circuit.

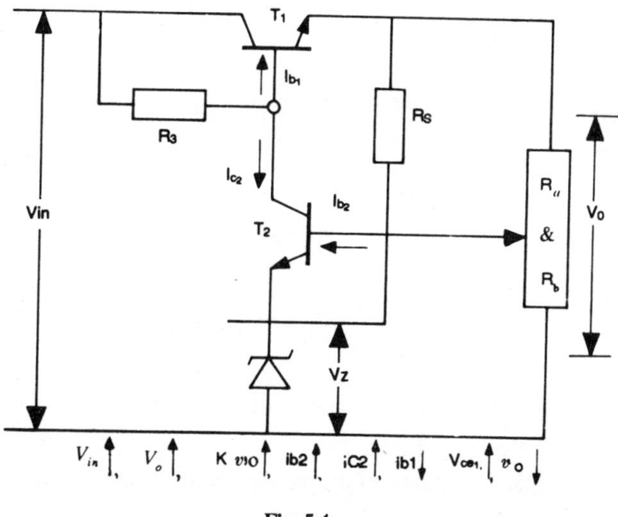

Fig. 5.4

Let us analyse the circuit for stabilisation and output resistance. (Lower-case letters are used for denoting the change in the values and the upper-case letters are used to indicate the *dc* values.). Transistor T_2 compares part of the output voltage, $k \times V_o$, with V_z, where $$k = R_b/(R_b + R_a) \tag{5.3}$$

Resistance R supplying current to zener diode could have been connected to V_{in} instead of V_o, but since V_o is going to remain constant, the changes in I_z, and hence, in V_z are virtually avoided. The flowchart for the regulating action is shown in the same figure.

As per this, if the input voltage, V_{in}, increases by v_{in}, the output voltage, V_o, will also increase proportionally by v_o. Part of this voltage, $k \times v_o$, will also increase, resulting in the increase of the base current I_{B2}, by i_{b2}. This base current is amplified and is available as i_{c2}. Thus, the change in the difference between V_o and V_z is amplified, and is available as i_{c2}. This will force a reduction in I_{B1} by i_{b1}. This may also be reasoned thus: If the input voltage increases, so will the output voltage for the transistor T_1. This output voltage is the emitter voltage for T_1. Also, the base voltage of T_1, being the collector voltage of T_2, will decrease since current I_{C2} has increased. Therefore, V_{BE1} will decrease, reducing the base current of T_1, and, hence, V_{CE1} will increase, effectively forcing V_o to go back towards its initial value and, thus, be forced to remain almost constant.

Let us derive the relation for voltage stabilisation. For finding out the change in the base current with the change in the output-voltage, let us Thevenise the T_2 base circuit. Refer to Fig 5.5

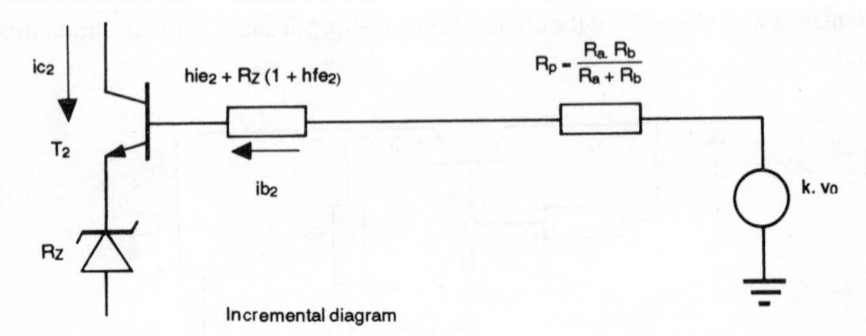

Fig. 5.5

we have $$i_{b2} = \frac{R_b \cdot v_o}{(R_a + R_b)} / (R_p + R_z\,(I + h_{fe2}) + h_{ie2}) \tag{5.4}$$

where $R_z\,(I + h_{fe2})$ is equivalent resistance of the zener resistance in the emitter circuit reflected in the base circuit.
and R_p is an equivalent resistance of the parallel combination of resistors R_a and R_b.

$$i_{c2} = h_{fe2} \cdot i_{b2}$$

$$= \frac{h_{fe2} \cdot k \cdot v_o}{R_p + h_{ie2} + R_z\,(1 + h_{fe2})} \tag{5.5}$$

Since current $I_3 = (V_{in} - V_o - V_{BE1})/R_3$

$$= I_{B1} + I_{C2} \tag{5.6}$$

The change produced in $I_3 = i_3$

$$= \frac{v_{in} - v_o}{R_3} \text{ (neglecting } v_{be1}\text{)}, \tag{5.7}$$

$$= \frac{v_{in}}{R_3} \text{ (assuming } v_o \text{ is very small)} \tag{5.8}$$

Assuming also that the change in I_L and, hence, the change in the base-current I_{B1} is negligible, since I_{B1} is approximately h_{FE1} times less than the load current, which is the collector current, we can write,

$$I_3 = I_{B1} + I_{C_2}$$

and

$$i_3 = i_{c2} + i_{b1}$$

$$= i_{c2} \tag{5.9}$$

From equations 5.5, 5.8 and 5.9, we can write,

$$\frac{v_{in}}{R_3} = \frac{h_{fe2} \times k \times v_o}{R_p + h_{ie2} + R_z\,(1 + h_{fe2})}$$

and, therefore, the voltage stabilisation factor can be given by,

$$S_v = \frac{v_o}{v_{in}} = \frac{R_p + h_{ie2} + R_z \cdot (1 + h_{fe2})}{R_3 \times h_{fe2} \times k} \tag{5.10}$$

R_o is output resistance, when v_{in} is zero.
Refer to Fig. 5.6.

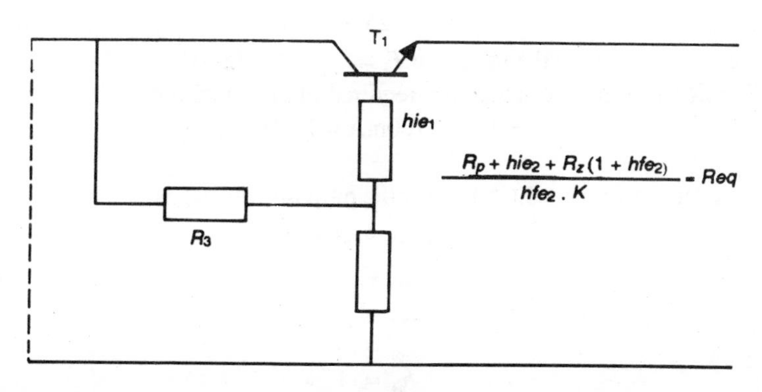

Fig. 5.6

$$R_o = \left[\left(\frac{R_{eq} \cdot R_3}{R_{eq} + R_3}\right) + h_{ie1}\right] \cdot \frac{1}{i + h_{fe1}}, \tag{5.11}$$

and if darlington pair is used, then

$$R_{od} = \frac{R_o + h_{ie3}}{1 + h_{fe2}} \tag{5.12}$$

These relations, at best, are only approximate, but they serve the purpose very well for all practical designs. We can deduce from equation 5.10 that the stabilisation will be better, i.e., the change in V_o will be less per unit change in V_{in}, if

(i) h_{fe2}, the current gain of the transistor T_2 is large, and

(ii) R_3 is very large, which is not possible since I_{C2}

and I_{B1} are to be supplied through the resistance R_3, and, therefore, the value of the resistance R_3 will be determined by these currents. However, we can employ a constant current source for current I_3, making the effective value of the resistance R_3 equal to that of the current source, namely, tending to infinity. From equations 5.11 and 5.12, it can be deduced that R_o will be less, if h_{fe1} is large. More often than not, the transistor T_1 is a power transistor having a fairly low value of the current-gain. We can use, if required, a darlington pair in place of a series pass transistor making gain of the combination to be equal to the product of the individual current-gains of the two transistors used in the darlington pair.

EXAMPLE 5.1

Design a voltage regulator circuit for 12V at 100mA. The circuit is to operate from a *dc* supply of 20 ± 5V.

Solution

1. *Selection of transistor T_1*

Voltage across the transistor $= V_{in(max)} - V_o$

$= 25 - 12 = 13.$

The current through the transistor $T_1 = I_{L(max)} = 100\text{mA}$.

Therefore, the power dissipation required of the transistor

$= 13\text{V} \times 100\text{mA} = 1.3$ Watts

We select a transistor ECN100 for the purpose, which has

$P_{d(max)} = 5\text{W at } 25°\text{C}.$

$I_{c(max)} = 0.7\text{A at } 25°\text{C}.$

$h_{FE(min)} = 50.$

$h_{fe(typ)} = 90.$

$h_{ie} = 1.3\text{k}$

The base current required = $I_{B1} = I_{L(max)} / h_{fe(min)}$
$= 100mA / 50$
$= 2mA.$

2. *Selection of the zener diode*
Since the output voltage is V_o = 12 Volts, we may select a zener diode of a voltage of 6.8V (between about 50 and 80% of the output voltage). The power dissipation capacity of the zener should be around 150mW giving,

$$R_s = \frac{V_o - V_z}{I_z}$$

$$= \frac{12 - 6.8}{5\,mA} = 1.04\text{k-Ohms.}$$

We select R_s = the nearest available standard value of 1k-Ohms, which should have power dissipation capability of $(5.2mA)^2 \times 1k = 27mW$.
We, therefore select R_s = 1k and of 1/8 watts.

3. *Selection of transistor T_2.*
As a thumb-rule, we select I_{C2} as about, between 10 and 50% of the I_{B1} depending upon the load current. Let I_{C2} = 1mA., in our case. The voltage across the transistor T_2 =

$$= V_{CE2} = V_o + V_{BE1} - V_z$$

$$= 12 + 0.6 - 6.8$$

$$= 5.8V.$$

This means that the $P_{d(max)}$ of the transistor T_2
$= 5.8V \times 1mA = 5.8mW$.
We select BC147B, which has,

$P_{d(max)} = 0.25W.$
$I_{c(max)} = 0.1A.$
$h_{FE(min)} = 200.$
$h_{fe(min)} = 240.$
$h_{ie} = 4.5\text{k-Ohms.}$

4. *Selection of the resistance R_3.*
The maximum base current required for the transistor for the maximum collector current of 100mA (which in reality is the load current and hence the emitter current). We neglect the additional current required for zener and the base current, making load current = emitter current = collector current.

$$= I_{C1} / h_{FE(min)} = 100\text{mA} / 50 = 2.0\text{mA}.$$

Also the value of $I_{C2} = 1$mA, giving the value of the current

passing through the resistance $R_3 = I \;=\; I_{C2} + I_{B1}$
$= 2\text{mA} + 1\text{mA} = 3\text{mA}.$

Minimum Voltage across the resistance $R_3 \;=\; V_{in(min)} - V_o + V_{BE1}$
$= 15 - 12 + 0.6$
$= 3.6$ Volts.

Hence,

$$R_3 = 3.6\text{V} / 3\text{mA}$$

$$= 453 \text{ Ohms.}$$

Selecting the nearest standard value available, $R_3 = 470$ Ohms, which will make the total current to $13.6 / 470 = 2.9$mA.
Since I_{B1} is dependent upon the load current and hence, will be 2mA, I_{C2} will reduce from 1.0mA to 0.9mA.
The power dissipation will be $(3\text{mA})^2 \times 4.7\text{k} = 39$mW.
We select $R_3 = 4.7$k and of 1/8W.
Likewise, when the input voltage is maximum, i.e., 25V, the current through the resistance R_3 will be equal to

$$(25 - 12 - 0.6)\text{V} / 470 \text{ Ohms} = 12.4 / 470$$

= about 27mA.

Since this current of 27mA (less 2mA for the I_{B1}) becomes the collector current for the transistor T_2, we have $P_{d(max)}$ for transistor $T_2 = 5.8\text{V} \times 25\text{mA}$ = 145mW.
This is well within the capacity of the transistor selected, which otherwise would have to be changed to a higher power transistor.

5. *Selection of R_a and R_b*

Voltage $V_{B2} \;=\; V_z + V_{BE2}$
$= 6.8 + 0.6$
$= 7.4$ Volts.

Therefore,

$$\frac{R_b}{R_a} = \frac{V_{B2}}{V_o} = \frac{7.4}{12} = 0.62,$$

provided that the value of R_b finally selected be sufficiently small as compared to $\{h_{IE2} + R_z(1 + h_{FE2})\}$, so that the shunting effect of the transistor T_2

is negligible, i.e., it is assumed that the current drawn by the R_a and R_b combination is very large as compared to the base current of transistor T_2, which is

$$= I_{b2} = 0.9 / 200 = 4.5 \text{ uA.}$$

If we select, I_r as about 1mA, (alternatively, I_r should be chosen, as a thumb-rule, to be about 10% of the load current), we get,

$$\begin{aligned} R_a + R_b &= V_o / I_r \\ &= 12 / 1\text{mA} \\ &= 12\text{k-Ohms.} \end{aligned}$$

and,

$$\frac{R_b}{R_a} = 0.62$$

Hence, $\quad R_a + 0.62\,R_a = 12\text{k-Ohms.}$

$$R_a = 7.4\text{k-Ohms.}$$

and, $\quad R_b = 4.6\text{k}$

We can select $R_a = 6.8\text{k}$, $R_b = 3.9\text{k}$ and connect a potentiometer of 1k in-between giving total resistance = 6.8 + 3.9k + 1k

$$= 11.7\text{k}$$

Adjusting the potentiometer, we will be able to adjust the output voltage finely to the required value.

The power dissipation for R_a $= (I_r)^2 \times 6.8k$ $= 6.8\text{mW.}$

Similarly, for R_b $= 4.7\text{mW}$

and for the potentiometer $= 1\text{mW}$

We may select all these components of 1/8 W capacity.

Let us calculate S_v and R_o for the above example.

From equation 5.10, we have

$$S_v = \frac{R_p + h_{ie2} + R_z\,(1 + h_{fe2})}{R_3 \times h_{fe2} \times k}$$

where $\quad R_p = 7.4\text{k}$ in parallel with 4.6k

$$= 2.8\text{k.}$$

R_z may be assumed as about 10 ohms, at the point of operation, and

$$k = (R_b)/(R_a + R_b)$$
$$= 0.383$$

Hence,
$$S_v = \frac{2.8k + 4.5k + 10\,(1 + 240)/1000}{0.47 \times 240 \times 0.383}$$
$$= 0.225$$

And from equation 5.10, we have,

$$R_o = 5.1 \text{ ohms.}$$

These are fairly large values.

We can improve upon these values, if so desired, by using a darlington pair of transistors in place of a single transistor T_1. This will help in increasing the value of the resistance R_3 since the base current requirements will be reduced, resulting in smaller value of the voltage-stabilisation ratio. Since the total gain increases the product of the individual current-gains of the transistors used in the pair, the value of the R_o will reduce to a considerable extent. This is demonstrated in the following example.

EXAMPLE 5.2

Design a series voltage regulator, using feed-back, with the following specifications.

$V_o = 12\text{V}$	$S_v < 0.03$
$I_L = 0.7$ to 1A	$R_o < 0.6$ Ohm

Solution

Since the output voltage required is 12V, the input voltage, V_{in}, can be assumed to be about $2 \times V_o$ = say, 25V.

Therefore, let
$$V_{in} = 25 \pm 5\text{V}.$$

i.e.,
$$V_{in(\min)} = 20\text{V} \text{ and } V_{in(\max)} = 30\text{V}.$$

1. The current flowing through the series loser transistor, T_1 is the load current. Note here that we neglect the current through R_3 and $(R_a + R_b)$ combination

$$I_{C1(\max)} = I_{L(\max)} = 1\text{A}$$
$$I_{C1(\min)} = I_{L(\min)} = 0.7\text{A}$$

Maximum voltage across the series loser transistor

$$V_{CE1(\max)} = V_{in(\max)} - V_o$$

$$= 30 - 12$$

Therefore, $\quad V_{CE1(max)} = 18$ volts.

Maximum power dissipated in the series loser transistor T_1

$$= V_{CE1(max)} \times I_{C1(max)}$$
$$= 18\text{V} \times 1\text{A}$$
$$= 18 \text{ Watts}$$

Allowing for temperature variation, we take the power rating of transistor T_1 about twice the calculated value.

Hence, $\quad P_{d1} = 18 \times 2 = 36$ Watts.

We select transistor ECN055 which has a power rating of 50W at 25°C and $I_{c(max)} = 5\text{A}$

V_{CEO} = 50 Volts which is greater than $V_{CE1(max)}$ required.

$h_{FE(min)} = 25.$

$h_{FE(max)} = 100.$

$h_{fe(typ)} = 75.$

Voltage at the emitter of transistor T_1 = $V_o = 12$ Volts

Voltage at the base of transistor $T_{,1}$ = $V_{E1} + V_{BE1}$

$$= 12 + 0.6$$
$$= 12.6 \text{ volts.}$$

2. The breakdown voltage of zener diode is not critical as long as it remains constant for normal operation. The diode is chosen with a low dynamic resistance, having slightly positive temperature coefficient. Usually, zener diodes having voltage-ratings > 6V provide the positive temperature coefficient, which will balance the negative temperature coefficient of transistor T_2. Let us select zener diode of, as mentioned in the last example, 6.2 Volts.

Now,

$I_{F1(max)} = 1\text{A}$

$h_{FE(min)} = 25$

therefore,

$$I_{B1(max)} = \frac{I_{C1(max)}}{h_{FE1(min)}} = \frac{1.0}{25} = 40 \text{ mA}$$

let $\quad I_{C2} = 0.2 \times I_{B1(max)}$

$$= 8\text{mA}.$$

$$I = I_{B1(max)} + I_{C2}$$

$$= 40 + 8 = 48 \text{ mA.}$$

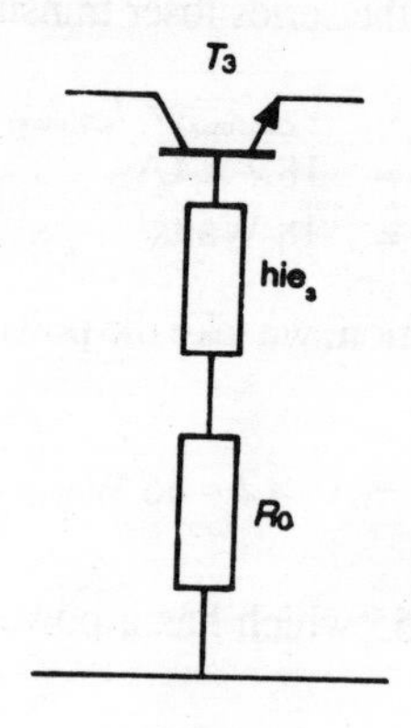

Fig. 5.7

From Fig 5.7, we have,

$$V_{in} = V_{R3} + V_{BE1} + V_o$$

therefore,

$$V_{R3(max)} = 25 - 0.6 - 12$$
$$= 12.4\text{V}$$

Therefore, $R_3 = \dfrac{V_{R3}}{I}$

$= 12.4 / 48\text{mA}$

$= 258$ Ohm

$= 270$ Ohms, the nearest standard value available.

Power dissipated in $R_3 = (48\text{mA})^2 \times 270$

$= 0.622$ Watts.

Hence, we select, $R_3 = 270$ Ohm, 1 Watt.

3. Maximum current through transistor T_2 occurs at $V_{in(max)}$ and $I_{L(min)}$

$V_{in(max)} = 25$ volts

$I_{C1(min)} = 0.7$A.

Therefore,

$$I_{B1(min)} = \frac{I_{C1(min)}}{h_{FE(max)}} = \frac{0.7}{100} = 7\text{mA.}$$

$$I_{C2(max)} = I_{3(max)} - I_{B1(min)}$$
$$= 48 - 7$$
$$= 41 \text{ mA.}$$

Since $I_{CQ} = 8 \text{ mA}$

and
$$V_{CE2} = V_o + V_{BE1} - V_z$$
$$= 12 + 0.6 - 6.2.$$
$$= 6.4 \text{ Volts}$$

We select transistor T_2 to be BC147A having $h_{FE(min)} = 115$ and $h_{ie} = 2.7\text{k}$.

4. Now,

$$V_{BE2} = 0.6 \text{ Volts}$$

Therefore,
$$V_{B2} = V_z + V_{BE2}$$
$$V_{B2} = 6.2 + 0.6$$
$$= 6.8 \text{ volts}$$

Also, $V_{B2} = k \times V_o$ where, k is as per equation 5.3

Therefore,

$$k = \frac{V_{B2}}{V_o} = \frac{6.8}{12}$$

$$= 0.567$$

i.e.,
$$k = \frac{R_b}{R_a + R_b} = 0.567 \quad \text{(i)}$$

Also
$$V_1 = V_o - V_{B2}$$
$$= 12 - 6.8 = 5.2\text{V}$$

Let the current through R_a and R_b be just about 1% of maximum load-current, say, 10 mA

Therefore
$$R_a = \frac{V_1}{10 \text{ mA}} = \frac{5.2\text{V}}{10 \text{ mA}}$$

i.e., $R_a = 520$ Ohm, having $P_d = (10\text{mA})^2 \times 520$.

and $R_b = 6.8\text{V} / 10\text{mA} = 680$ Ohms

with $P_d = (10\text{mA})^2 \times 680$

To facilitate fine adjustment of the output voltage, we select

$$R_a = 470 \text{ ohm, 1/8 watt.}$$
$$R_b = 620 \text{ ohm, 1/8 watt.}$$

and a potentiometer of 500 ohms in the centre.

5. *Calculation of S_v* Using (5 – 10), we get,

$$S_v = \frac{294.7 + 2700 + 10\,(1 + 115)}{270 \times 115 \times 0.567}$$

assuming R_z about 10 ohm

$$S_v = 0.236$$

Since S_v required = 0.03 or less, we have to go in for a darlington-pair connection as this will permit us to increase the value of the resistance R_3.

6. *With Darlington connection*
Selection of T_1 remains the same.
i.e., The transistor T_1 = ECN055.

7. *Selection of transistor T_3.*

$$\begin{aligned} V_{CE3(max)} &= V_{in(max)} - V_o - V_{BE1} \\ &= 25 - 12 - 0.6 \\ &= 12.4 \text{ volts.} \\ I_{C3} = I_{E3} &= I_{B1(max)} = 40 \text{ mA} \\ P_{d3(max)} &= V_{CE3(max)} \times I_{C3} \\ &= (12.4\text{V}) \times (40 \text{ mA}) \\ &= 0.496 \text{ W.} \end{aligned}$$

We select T_3 as ECN100 having the following specifications:

$$\begin{aligned} P_{d(max)} &= 5\text{W at } 25°\text{C}, \\ I_{C(max)} &= 700 \text{ mA}, \\ V_{CEo} &= 60\text{V}, \\ h_{FE(min)} &= 50, \\ h_{fe(typ)} &= 90. \end{aligned}$$

$$I_{B3(max)} = \frac{I_{C3}}{1 + h_{FE3(min)}} = \frac{40}{1 + 50} = 0.8 \text{ mA.}$$

This will permit us to select a smaller value of I_{C2}.

Let $$I_{C2} = 2\text{mA}$$

8. *Selection of R_3*

$$R_3 = \frac{V_{in(min)} - V_o - V_{BE1} - V_{BE3}}{(I_{C2} + I_{B3})}$$

$$= (15 - 12 - 0.6 - 0.6) / 2.8\text{mA}$$

$$= 4.57\text{k}.$$

We select the resistance R_3 = 4.7k-Ohms, being the nearest standard value available. The power dissipation capacity should be 1/4W since the actual power dissipation is

$$= (2.8\text{mA})^2 \times 4.7\text{k} = \text{about } 37\text{mW}.$$

9. Maximum current through the resistance R_3

$$I_3 = \frac{V_{in(max)} - V_o - V_{BE1} - V_{BE3}}{4.7\text{k}}$$

$$= \frac{25 - 12 - 0.6 - 0.6}{4.7\text{k}}$$

$$= 2.51\text{mA}.$$

$$I_{b1(min)} = I_{L(min)} / h_{FE(max)}$$

$$= 0.7/100 = 7\text{mA}.$$

This current is being supplied by the transistor T_3 and hence, is the collector current for the transistor T_3. The minimum base-current for the transistor T_3 = $I_{B3(min)}$

$$= I_{B1(min)} / h_{FE3(max)} = 7\text{mA} / 280$$
$$= 25 \text{ uA}.$$

Since we have assumed that the I_{C2} = 2mA,

therefore,

$$I_{2(max)} = I_{(max)} - I_{3(max)}$$
$$= 2.51\text{mA} - .025\text{mA}$$
$$= 2.485\text{mA, say, } 2.5\text{mA}.$$

and

$$I_{C2(min)} = 2\text{mA}.$$

$P_{d(max)}$ for the transistor T_2

$$= (V_{B3} - V_z) \times I_{C2(max)}$$
$$= \{(12 - 0.6 - 0.6) - 6.2\} \times 2.5\text{mA}.$$
$$= 11.5\text{mW}$$

The transistor T_2 can be the same BC147A, which we had selected earlier. The value of S_v will now be equal to

$$S_v = \frac{294.7 + 2700 + 10\,(1 + 115)}{4700 \times 115 \times 0.567}$$

$$= 0.014$$

which is better than the required value of 0.03.

From equation 5.11, we have the value of $R_o = 0.183$, which is less than, and better than, the value 0.6 Ohm.

EXAMPLE 5.3

For Example 5.2, design a preregulator circuit, over-load protection circuit and heat-sink, if necessary.

Solution

Pre-regulator Circuit

As seen in the previous example, we could increase the value of the resistance R_3 from 470 ohms to 4.7k-Ohms, by using a darlington pair. This increase in the value of the resistance, improved the voltage stabilisation factor from 0.236 to 0.014. This can be improved even further by using a constant current source in place of the resistance R_3. (Refer to Fig. 5.8)

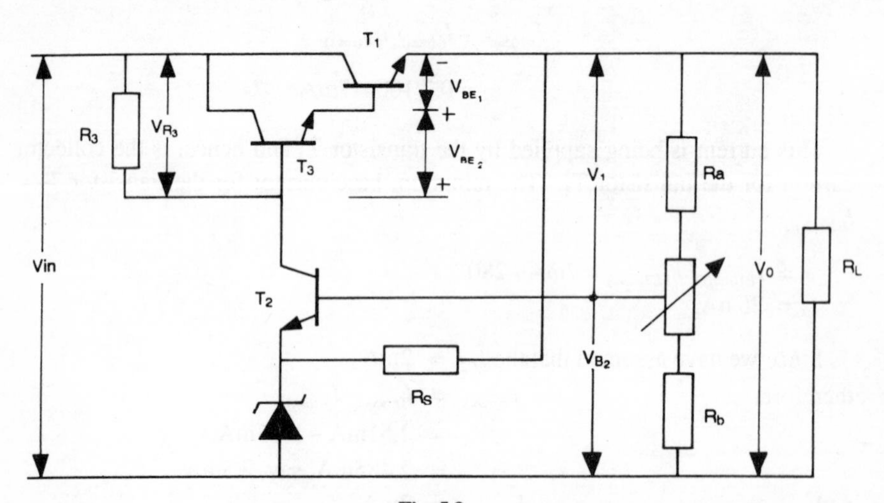

Fig. 5.8

Since D_2 is a zener diode, it will keep the voltage between the base of the transistor T_4 and the positive supply rail reasonably constant. This would also keep the voltage across the resistance R constant, since V_{BE4} is also virtually constant.

Hence, V_R, the voltage across the resistance R which

$= V_{z2} - V_{BE4}$ will remain constant, and

so will be the current I_3, being $= V_R / R$.

This circuit replaces the resistance R_3 of the previous circuit. The dynamic value of the resistance R_3, therefore, $= v_{in}/i_3 = \infty$, since i_3, the change in the current I_3 is zero.

The value of the S_v becomes, practically, a very small value, making v_o, the change in the output voltage V_o

$= S_v \times v_{in} =$ a few milli-volts.

For our circuit we need,

$$I_3 = 2.51 \text{ mA.}$$

The transistor T_4 required is a PNP transistor. We can select either BC157A or 2N525, both are small signal transistors and easily available. We select 2N525 which has $h_{FE(min)} = 35$. Since the current required $I_{E4} = I_{C4} = 2.51$mA, we will need the base-current for the transistor T_4 as = 2.51mA / 35 = 71.4uA.

We select a zener diode of 4.7V. (for no specific reason, except that it should be of voltage less than 6V, which should, normally, provide negative temperature co-efficient to overcome the negative temperature co-efficient of the transistor T_4.) We shall operate the zener at 5mA, giving us,

$$R_5 = \frac{V_{in(min)} - V_{z2}}{5\text{mA}}$$

$$= \frac{20 - 4.7}{5\text{mA}}$$

$$= 3.06\text{k}.$$

We select 3.33k, being the nearest available standard value. Recalculating, we get I_{z2} = (20 – 4.7)V / 3.33k = 4.64mA.

$$\begin{aligned} \text{Resistance } R &= (V_R) / I_3 \\ &= 4\text{V} / 2.51\text{mA} \\ &= 1.6\text{k}. \end{aligned}$$

We may select a resistance of 1k in series with a variable resistance of 1k.

Heat-Sink calculation

Let the ambient temperature be 30°C and the case temperature $T_{c(max)}$ = 90°C. For the power transistor ECN055 that we selected, we have, (from the manufacturer's data-sheets)

$$T_{j(max)} = 200°\text{C}.$$

$$\theta_{J-C} = 3.5°\text{C} / \text{W}.$$

The maximum power dissipation for the transistor is going to be 18W. Taking a slightly higher value, we shall assume power dissipation of, say, 20W. Refer to Fig 5.9.

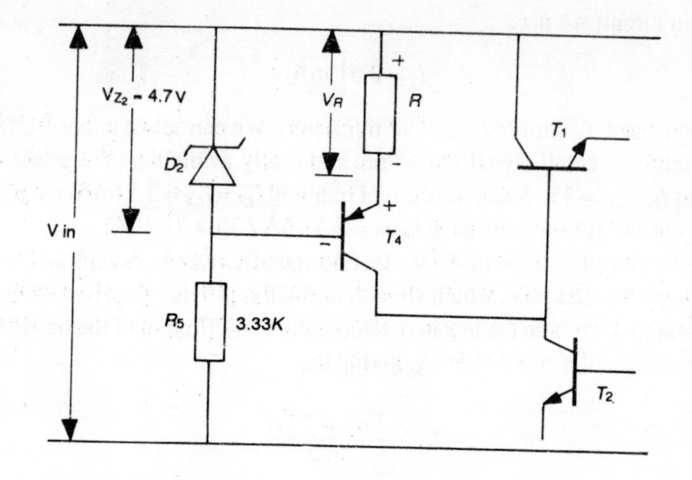

Fig. 5.9

Normally, since there is a mica wafer between the case of the transistor and the heat-sink for electrical isolation, the thermal resistance is assumed to be about

$$\theta_{c-s} = 0.75°C/W$$

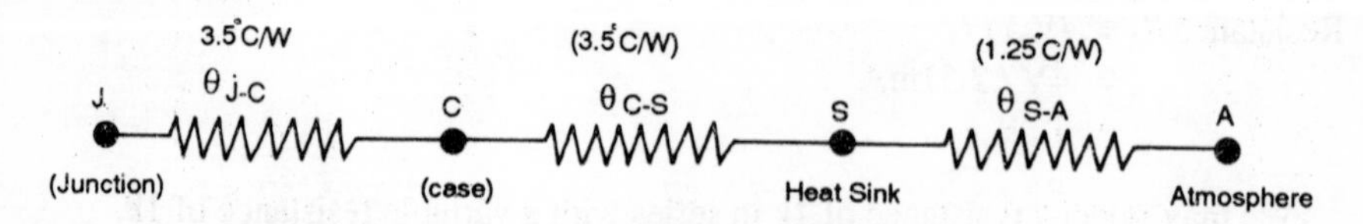

Fig. 5.10

Since, the power to be dissipated is 20W, we have,

$$\theta_{C-A} = (T_C - T_A)/P_d$$

$$= (90 - 70)/20$$

$$= 2°C / W$$

But $$\theta_{C-A} = \theta_{C-S} - \theta_{S-A}$$

Therefore, $$\theta_{S-A} = 2 - 0.75$$

$$= 1.25°C /W$$

We shall have to select a heat-sink having a thermal resistance = θ_{S-A} = or, less than, 1.25° C/W.

Let us check for the junction temperature.

$$P_d = \frac{T_j - T_c}{\theta_{J-C}},$$

Hence, $T_j = 90 + 20 \times 3.5$

$= 160°C$, which is within the specified value.

6

CONTROLLED RECTIFIERS

6.1 Drawback of Unregulated and Transistorised Regulated Supplies

In the previous chapters two distinct characteristics were noted:

(a) When the diodes are used for rectification, there is a definite relation between the *dc* voltage available and the *ac* voltage required, which has to be obtained through a transformer having a proper turns ratio. If we need a *dc* voltage less than for what it is designed, we may have to use a series resistance to reduce the voltage.

(b) In case of the series-transistor voltage regulators, the above difficulty is overcome to a great extent as one is able to change the output voltage by simply changing the resistors R_a and R_b. But again this has the limitation that the output voltage available is restricted by the V_{CEO} of the transistor used. The power rating of the transistor used, as a series loser, depends on the voltage difference between the input and the output voltages.

6.2 Silicon Controlled Rectifier

Both these difficulties are minimised to a great extent by using thyristors. The term, **Thyristor** applies to a family of multilayer semiconductor devices that exhibit two states, ON-OFF, both stable, due to their inherent regenerative feedback. These devices have the ability to control large power with minimal control power, and this control power does not have to remain continuous as in case of the base-current of a transistor. For instance, a transistor of, say, 50A may need about 1 to 1.2A of continuous base current at about 0.8V, needing about 1W of continuous power, whereas a thyristor of the same capacity may need a momentary pulse of about 50mA at about 1.2V, needing about 0.6 W of power, and that too, for a short duration. Apart from this, transistors for a relatively high-voltage high-power operations are rarely, if at all, available.

These devices can be used where current, and hence, the power required vary from small values to several hundreds of amperes. We shall be dealing here with the use of one of these devices, namely SCR, in circuits for rectification to *dc*. The SCR—Silicon Controlled Rectifier—is a member of the thyristor family and is unidirectional, three-terminal, reverse-blocking device. To that extent it is like a normal rectifier diode. The difference is that SCR blocks in the forward direction also, until it is turned-ON.
Refer to Fig. 6.1.

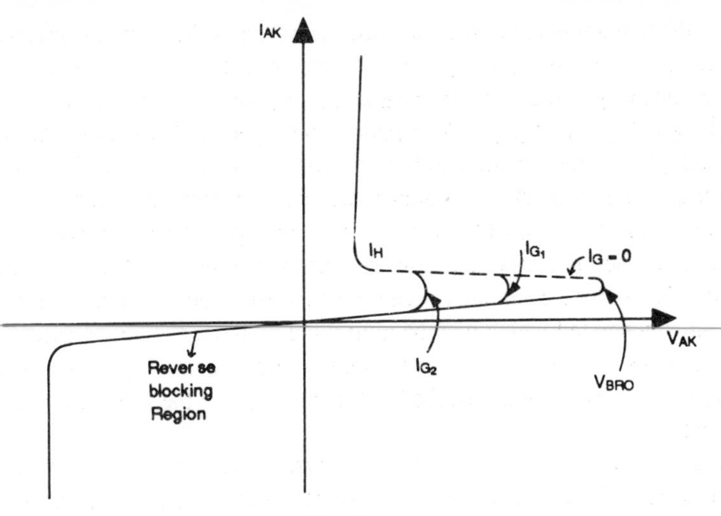

Fig. 6.1

The characteristic shown is a plot of the anode-current, I_{AK} against anode to cathode voltage, V_{AK}, with the gate circuit open. When the anode voltage is increased, the SCR blocks the current until the voltage exceeds the forward-breakover voltage, V_{BRO}. At this point the SCR changes from non-conducting to high conduction state. The anode current, now, is limited only by the external load resistance. To turn the SCR OFF from this high-conduction state, the only way is to reduce the anode current to a value below a minimum current, called the *holding-current*. Increasing the gate current reduces the forward-breakover voltage considerably. This means that the SCR can be turned-ON at relatively small voltages by sending in appropriate amount of the gate-current.

6.3 Selection Criterion for SCR

In our applications, we shall select SCR having a value of V_{BRO}, - the forward-breakover—voltage when the gate current is zero (strictly speaking, when gate circuit is open)—much larger than the normal operating voltage required. For example, when SCR is supposed to be working with the applied voltage, of 230 volts rms, the forward breakover voltage should be greater than the peak value of the supply voltage, i.e., it should be greater than, $\sqrt{2} \times 230 = 325\text{V}$. Likewise, the reverse breakover voltage should be greater than −325 volts. A more practical value should be 400 volts to allow for the fluctuations in the supply voltage. Yet another problem that the designer faces is the excessive leakage current flowing when negative voltage is applied across the device, and at the same time if the gate voltage is applied. Normally, therefore, the gate voltage is withdrawn when the negative anode voltage is present.

The junction temperature causes yet another problem. Since SCR is normally used in power control applications, the dissipation should be carefully looked into.

$T_{j(max)}$, the maximum junction temperature at which the SCR operation is permitted, is perhaps one of the most important parameter of the SCR, since almost all other ratings, one way or the other, are related to it. For instance, V_{DRM}, *the peak off-state blocking voltage*, or V_{DROM}, *the repetitive peak forward off-state voltage*, refers, respectively, to maximum anode-to-cathode voltage that can be applied and maximum anode-to-cathode voltage that can be applied **repeatedly**, without triggering the SCR into ON state. These voltage ratings are specified at the maximum junction temperature rating, since these represent the worst-case condition. This is done essentially because at this temperature, the maximum leakage current is generated which may be multiplied by the regenerative action of SCR, turning it ON. This current, when it crosses the gate-cathode junction, reduces the forward break-over voltage of the SCR, behaving as if the gate current is actually flowing.

Since we will be considering the use of the SCR in essentially low frequency applications - the power frequencies - we need not dwell more on the gate circuit limitations and characteristics, except that we should take care about the sensitive SCRs. These SCRs need very low gate current to trigger. Hence, where such SCRs are used, care should be taken to see that during the off-state, gate circuit does not present any leakage current, like in case of directly connected triggering circuit. In such cases, we use devices like pulse transformers for proper isolation, or a small resistance in series with the gate. Again, the resistance so connected should be carefully selected so as not to produce voltage which crosses the limit specified by the manufacturer.

6.4 Controlled Rectifier

Though these devices are basically switches, they can be used in linear applications like rectification, when we consider that the *dc* value of the voltage is an **average** value taken over a certain length of time.
Refer to Fig 6.2.

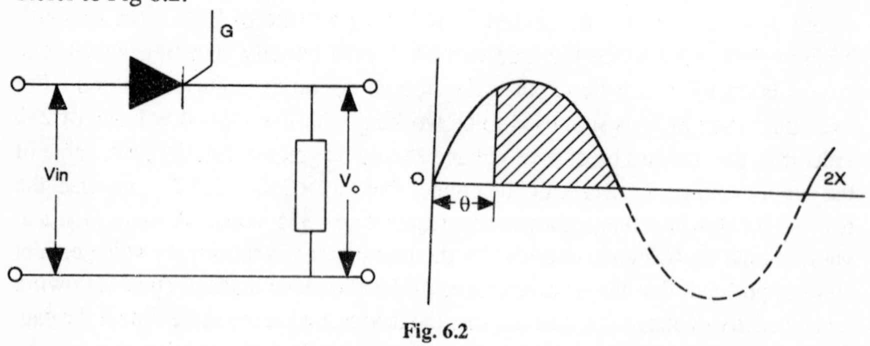

Fig. 6.2

The SCR will block not only the negative part of the applied sinusoidal voltage, but will also block the part of the positive waveform, upto a point SCR is triggered ON. This gives us a half-wave rectification if the angle θ is zero. For any other value of this **firing angle**, only *part* of the positive half-wave will be available.

Let us calculate the average voltage available. As in Chapter I, we can integrate the wave-form from its firing point upto the end of the positive wave-form. We get,

$$V_{av} = \frac{1}{2\pi}\int_{\theta}^{\pi} V_p \sin \omega t \cdot d\omega t$$

$$= \frac{1}{2\pi}\cdot [-V_p \cos t]_{\theta}^{\pi}$$

$$= (V_p/2\pi)\cdot(1+\cos\theta) \qquad (6.1)$$

$$= (V_p/\pi), \text{ if } \theta = 0.$$

Likewise,

$$V_{RMS} = \sqrt{\frac{1}{2\pi}\int_{\theta}^{\pi} (V_p \sin \omega t \cdot d\,\omega t)^2}$$

$$= \left[\frac{V_p^2}{4\cdot\pi}\int_{\theta}^{2\pi} (1-\cos 2\,\omega t)\cdot d\,\omega t\right]^{\frac{1}{2}}$$

$$= \left[\frac{V_p^2}{4\cdot\pi}[t - \sin 2\,\omega t/2]_{\theta}^{\pi}\right]^{\frac{1}{2}}$$

$$= (V_p/2)\sqrt{\frac{1}{\pi}[\pi-\theta+\sin 2\theta]} \qquad (6.2)$$

$$= (V_p/2) \text{ if } \theta = 0$$

6.5 Full-Wave Controlled Rectifiers

Shown in Fig. 6.3 are various methods of using SCRs in rectifying circuits. Fig. 6.3(a), shows a half-wave rectifier circuit. It can be seen that the half-wave circuit of Chapter I is simply redrawn, replacing a diode of that circuit by SCR. Circuit (b) is, likewise, a full-wave rectifier circuit, again replacing the two diodes by SCRs. A little modification of this circuit is shown in the circuit (c), where a SCR is placed in series to the load resistance. This gives a similar result, but provides

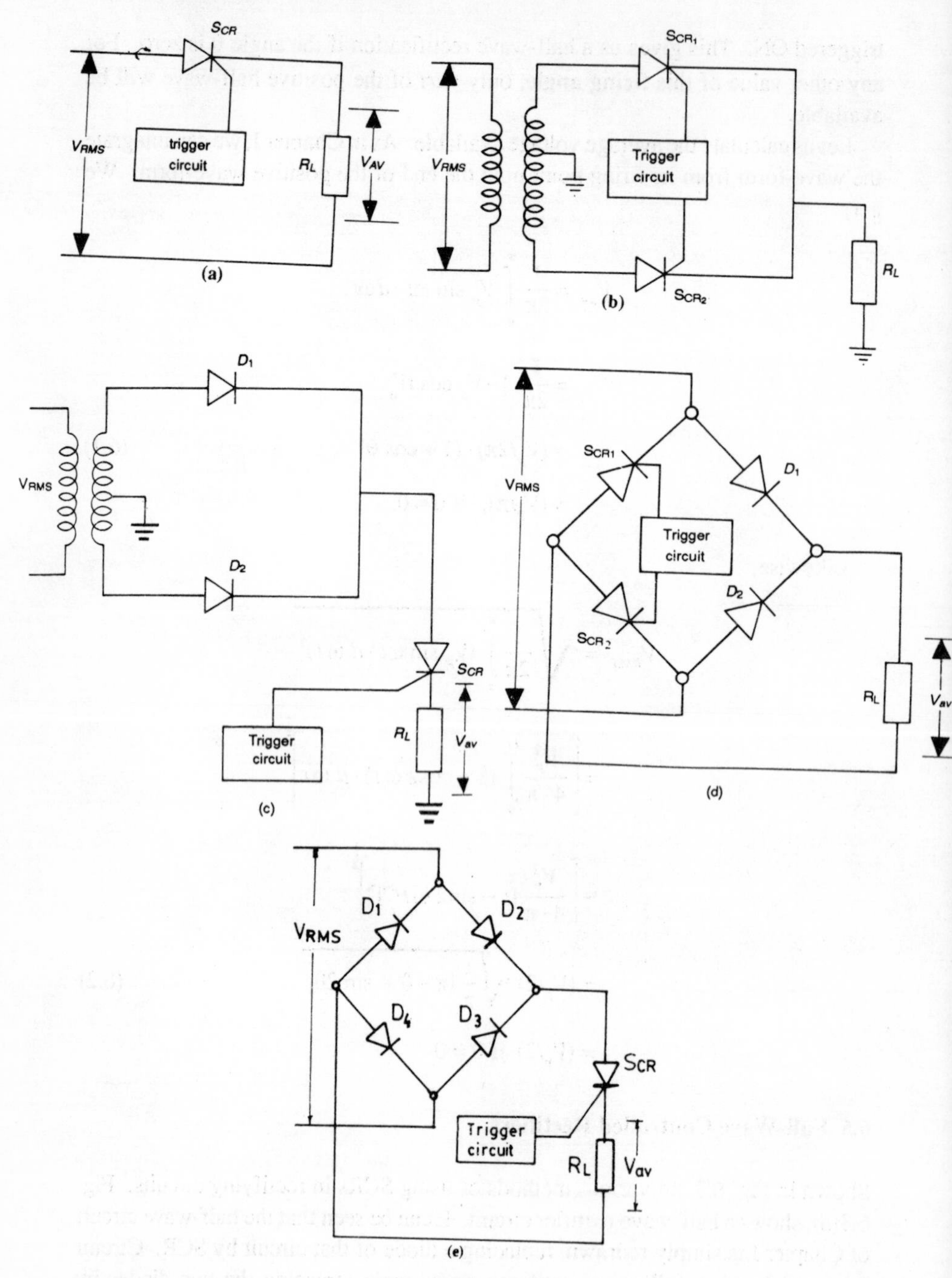

Fig. 6.3 (a, b, c, d and e)

simplification for the trigger circuit. Of course, it will need two diodes and one SCR. At high current applications this provides an economical solution, since a diode is cheaper than a SCR of similar rating.

Figure 6.3(d) is a bridge configuration for the rectifier circuit. This circuit employs two SCRs and two diodes in place of four diodes. The (e) part of the circuit depicts the same bridge configuration, but with a solitary SCR placed in series with the load resistance. Barring the center-tapped transformer configuration, no other circuit of rectification employing SCR/s needs a transformer since, at least theoretically, any voltage can be made available by simply changing the firing-angle of the SCR/s.

6.6 Triggering Methods

Let us now consider the triggering methods and the devices that can be used for the purpose. The triggering of the SCR can be achieved by a simple resistance and capacitance combination, as shown in Fig 6.4.

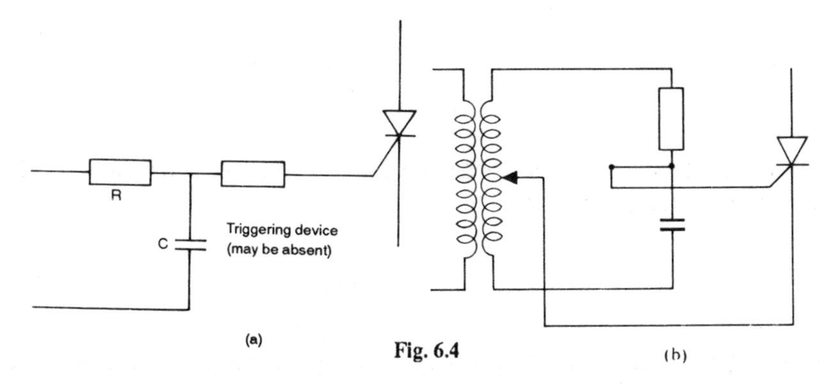

Fig. 6.4

These circuits, however depend heavily on the firing characteristics of a particular SCR used. Besides, the power level required by these control circuits is high because the entire triggering current has to flow through the resistance. In such circuits, change in the firing angle is not very inconvenient, but it becomes very difficult to introduce a correction automatically in case of a regulating system employing error feedback.

Pulse triggering can obviate most of these disadvantages. Since the pulse exists but for a very short duration, it can be used to overdrive the gate circuit, thereby taking care of the tolerences of various SCRs. Unlike R-C combinations alone, in pulse-circuits the charge can be stored in the capacitor and can be discharged instantly and more precisely. Since the time-constant for storing can be very large, relatively, the resistance used for the purpose can also be fairly large, resulting in low power levels.

6.7 Uni-Junction Transistor

We shall consider here Uni-Junction Transistor (UJT). The UJT, is a three terminal device. These terminals are called, the emitter E, the base-one (B_1) and the

base-two (B_2). Between the two bases, the transistor offers a simple resistance, R_{BB}. This is called the inter-base resistance and has a typical value of about 4.5K to about 9K.
Refer to Fig 6.5.

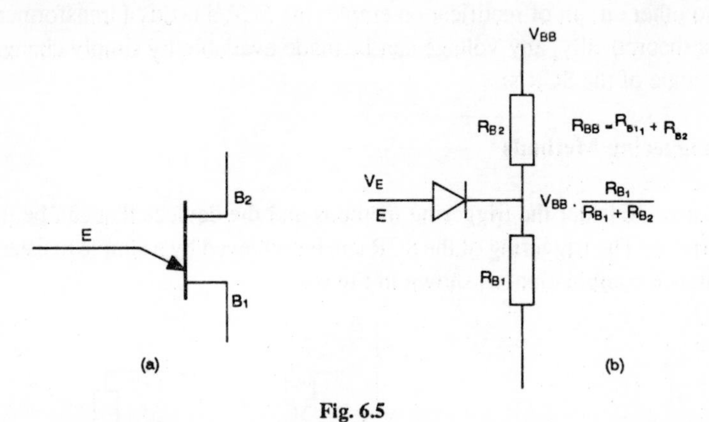

Fig. 6.5

If we consider the emitter as anode of a diode, the voltage at the junction will be a cathode voltage, and be equal to

$$V_{b1} = \frac{R_{b1}}{R_{b1} + R_{b2}} \times V_{BB} \tag{6.3}$$

If the voltage at the emitter is less than the voltage at the junction point, the diode is reversed biased, and hence only the leakage current, I_{EO}, will flow. Refer to Fig. 6.6.

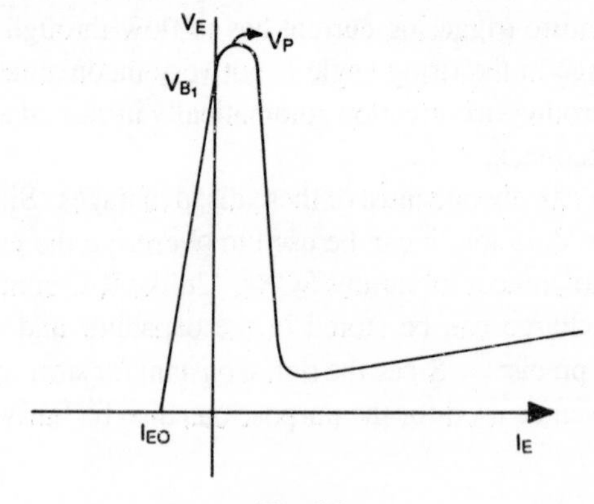

Fig. 6.6

As the voltage at the emitter is increased, the net reverse voltage at the emitter-base diode will go on decreasing, resulting in the decrease in I_{EO}, the leakage current, until at a point in time when the emitter voltage becomes equal to V_{b1}, this current ceases to flow through this diode.

If now the emitter voltage is increased beyond this point, the diode starts conducting when the voltage becomes one diode drop more than V_{b1} and the emitter current becomes slightly more than the I_p. This value of the emitter voltage is called the peak-voltage, V_p. From this point onwards the characteristic shows a very interesting development. The characteristic now enters into a negative resistance region. The resistance between the emitter and the base-one is now reduced to a very small value, and the current primarily is limited only by an external resistance connected to the emitter and base-one circuit. It should be noted that the peak-voltage of the UJT is dependent on the interbase resistance. Thus,

$$V_p = \eta \cdot V_{BB} + V_d \tag{6.4}$$

where, η is called the **intrinsic stand-off ratio.** The value of this ratio is, typically, between 0.5 and 0.8 at 25°C.

At this point, it will be necessary to introduce the effect of the temperature on the peak-voltage. Since the peak-voltage depends on the interbase resistance and the diode voltage drop, it is obvious that it will be very temperature sensitive. V_p, the peak-voltage exhibits a change of the order of between – 2mV and – 3mV / °C of the temperature increase, depending on the transistor used, this change can be entirely attributed to the change in the diode voltage, V_d. It is possible to compensate for this change in V_p if we keep in mind that the interbase resistance has a positive temperature co-efficient. If we introduce a resistance R_2 in series with the base-two as shown in Fig 6.7, the increase in temperature will cause the interbase resistance to increase, which, in-turn, will increase the voltage V_{BB} which is across R_{BB}. If the value of the resistance R_2 is properly chosen, it will, thus, permit the increase of V_{BB}, and provide correct compensation for the decrease in the diode voltage drop.

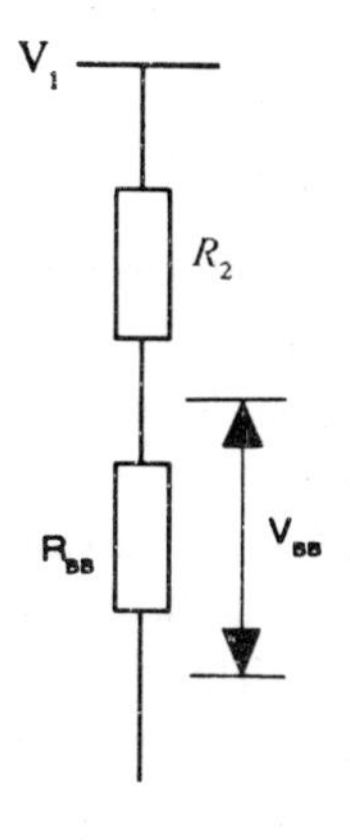

Fig. 6.7

Empirical formulae have been developed for the selection of the resistance R_2. These formulae vary a little for different types of the transistors used. In general, however, these formulae provide the value of the resistance R_2, to a fairly acceptable level of accuracy.

$$R_2 = \frac{10000}{\eta \cdot V_s} \tag{6.5}$$

Also,

$$V_{s(\min)} = \frac{(2200 + R_{b2})}{2300} \cdot V_s \tag{6.6}$$

6.8 UJT as a Relaxation Oscillator

Let us discuss the use of the UJT as a relaxation oscillator.
Refer to Fig 6.8.

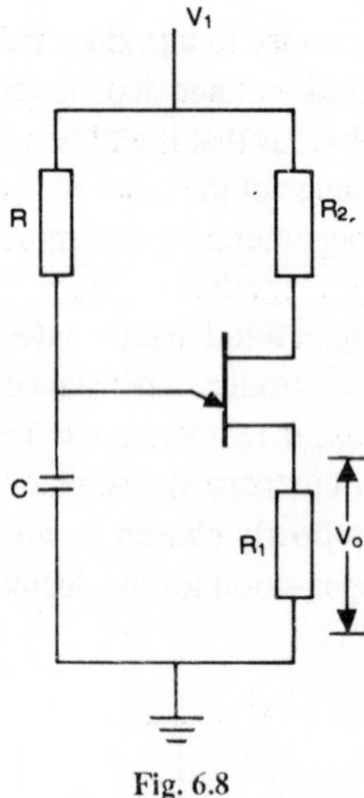

Fig. 6.8

The condenser, C, charges through the resistance, R. When voltage across the capacitor becomes equal to V_p, UJT will turn ON. Of course, whether the UJT can turn ON or not, will depend upon the value of the resistance R. Let us take a maximum value of the resistance R which will still permit the circuit to function as an oscillator. The load line for this maximum value of the resistance, R_{max}, is shown in Fig 6.9. Joining point V_1 to the vertical part of the device characteristic at a point V_p will produce this line. This line intersects the characteristics at a point **a**. As soon as the capacitor voltage reaches the value of the peak-voltage V_p, the UJT is turned ON. The condenser is now offered a very low resistance path, emitter to base 1, which will force the condenser to discharge almost completely. The time it needs for discharge is virtually decided by external resistance, R_1 connected in the B_1 lead. The operating point slides from **a** to **b**. In the process, the condenser is discharged by a peak pulse current, i_r, and produces a peak voltage pulse, across the resistance R_1. The magnitude of the voltage pulse produced will depend on the resistance R_1 and the gate impedence. The condenser voltage follows the characteristic from point **b** to point **c**, where the device negative resistance characteristic is tangential to the load line corresponding to resistance R_1, whereupon, the operating point gets transferred to point **d**. The condenser again starts charging through the resistance R.

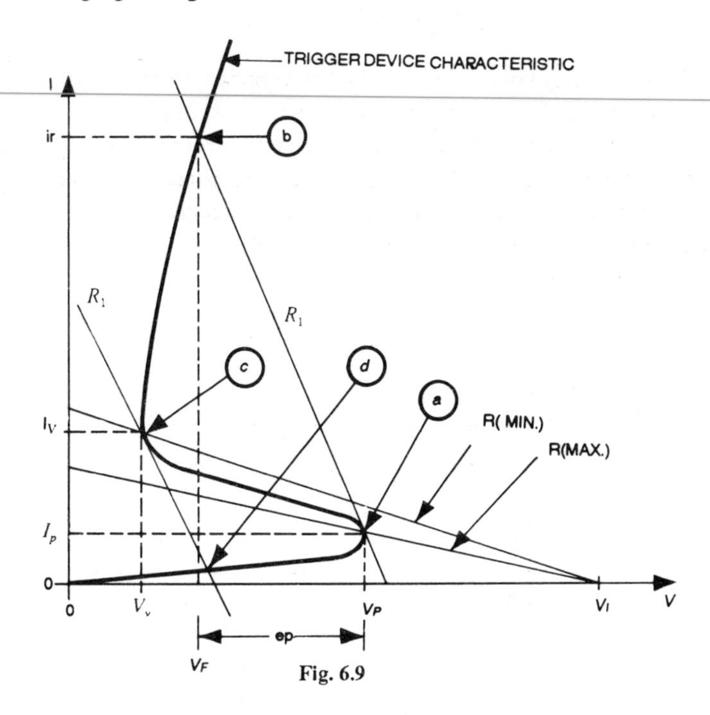

Fig. 6.9

Suppose, now, we reduce the value of the resistance R to a minimum possible value and yet expect the circuit to work as a relaxation oscillator. The load line corresponding to this value of the resistance R_{min} should pass through the point **c**. Any smaller value than this will cause the device to remain conducting somewhere between points **b** and **c**. Likewise, increasing the value of the resistance R to more than the oscillating value, R_{max}, will force the operating point between point **a** and the origin, effectively ceasing the oscillations.

One more point, though not very important from fairly low frequency operations' point of view, is that if the device operating point movement from point **a** to point **b** is not fast enough, it will never reach the point **b**, since the capacitor voltage itself is decreasing. The discharge time constant, therefore, should be very large as compared to the device switching time.

If we do not consider the switching time of the device of any significance, then the peak value of the voltage available across the resistance R_1 is equal to the difference between V_P and V_F. The magnitude of the current pulse will depend on the intersection point of the load line R_1 and the characteristic curve.

6.9 Frequency of Oscillations

The UJT used as a relaxation oscillator for triggering SCR will generate a pulse frequency which is dependent on, essentially, the values of the resistance, R, and the value of the capacitor, C.

The voltage across the capacitor is given by,

$$V_c = V_1 - (V_1 - V_v) \cdot e^{-t/CR} \tag{6.7}$$

where V_v is the valley voltage below which the condenser cannot discharge. Hence the condenser will always remain charged to this potential.

When this voltage reaches the value V_p, the UJT turns ON. As explained earlier, the capacitor will discharge through a low value resistance, R_1, producing a sharp voltage pulse. Let us assume that this happens at a time $t = t_1$.
Hence, at time $t = t_1$, we have, (Refer to Fig 6.10)

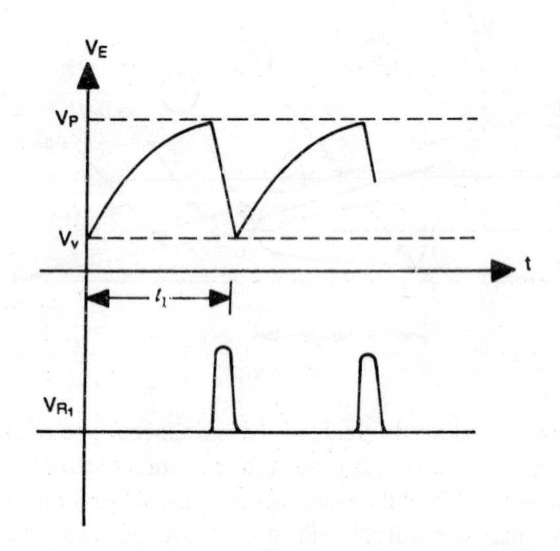

Fig. 6.10

$$V_p = V_1 - (V_1 - V_v) \cdot e^{-t/CR} \tag{6.8}$$

Again, since $V_p = \eta \cdot V_{BB}$, neglecting the diode drop V_d, substituting this value and rearranging, we have,

$$e^{-t/CR} = \frac{V_1 - \eta \cdot V_{BB}}{V_1 - V_v} \tag{6.9}$$

Further, assuming that the difference between V_1 and V_{BB} is small enough to be neglected, and that the value of the voltage V_v, is also negligible as compared to the voltage V_{BB}, equation 6.9 will reduce to

$$t = CR \ln \frac{1}{1-\eta} \tag{6.10}$$

If we neglect the discharge time of the capacitor, then the inverse of Equation 6.10 will give the frequency of oscillations.

6.10 Pulse Transformer

While discussing SCRs, it was mentioned that a pulse transformer may be used for providing electrical isolation between the trigger pulse generator and the SCR gate circuit. Where used, the pulse transformer has a turns ratio of 1:1, in cases where a single SCR is to be triggered, (refer to Figs 6.3 a, c, e) or has three windings of 1:1:1 turns ratio where two reverse connected SCRs are to be triggered. The secondary of the transformer is normally connected directly to the gate as shown in Fig 6.11.

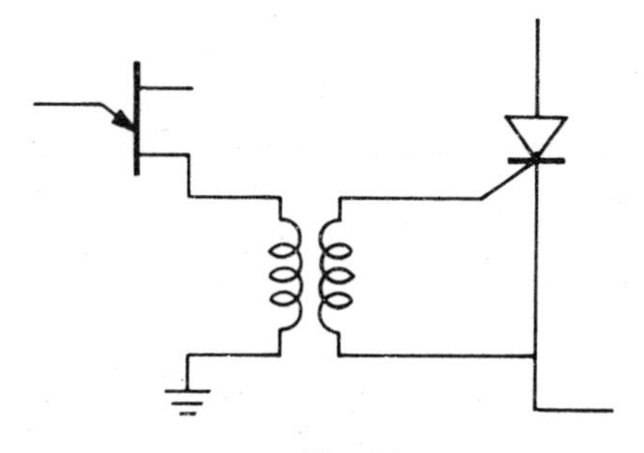

Fig. 6.11

Sometimes, however, a series-resistance may have to be added for reasons explained earlier, or a diode may be added to block the reversal voltage in the transformer secondary. In an electrically noisy environment, perhaps, the SCR may trigger due to the spurious spikes present. These can be avoided by using filter circuits, or in most cases loading the secondary by connecting a resistance across it. The drive current may have to be augmented as a result of this. It need not be emphasised here, that the pulse transformer should be of good quality and properly designed.

6.11 Synchronisation

Since the trigger circuit and the SCR circuit work independently, it is obvious that the trigger circuit will not be able to deliver the firing pulse exactly at the same instant in each cycle. If it is so, then the SCR will conduct for different intervals of time in different cycles.

There are two ways in which **synchronisation** can be provided in such circuits, viz.

1. The UJT lends itself to turning ON by accepting negative going pulse at the base-2 terminal, as shown in Fig 6.12.

This pulse will reduce the interbase voltage, effectively decreasing V_p, which will turn ON the UJT at any desired point in the cycle since momentarily the voltage V_{b1} will become less than the voltage V_E.

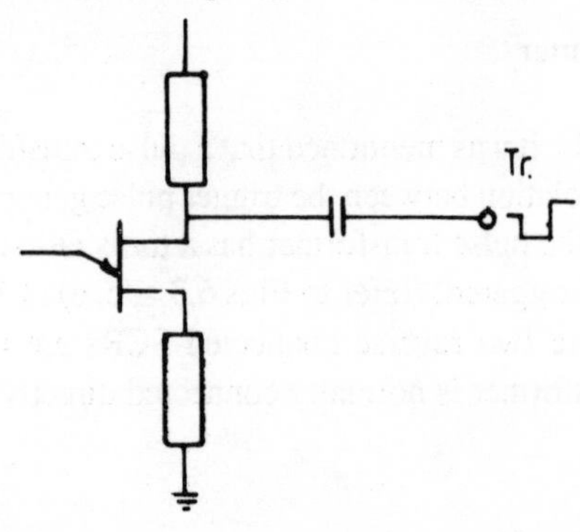

Fig. 6.12

2. The other method of achieving synchronisation is shown in Fig. 6.13.

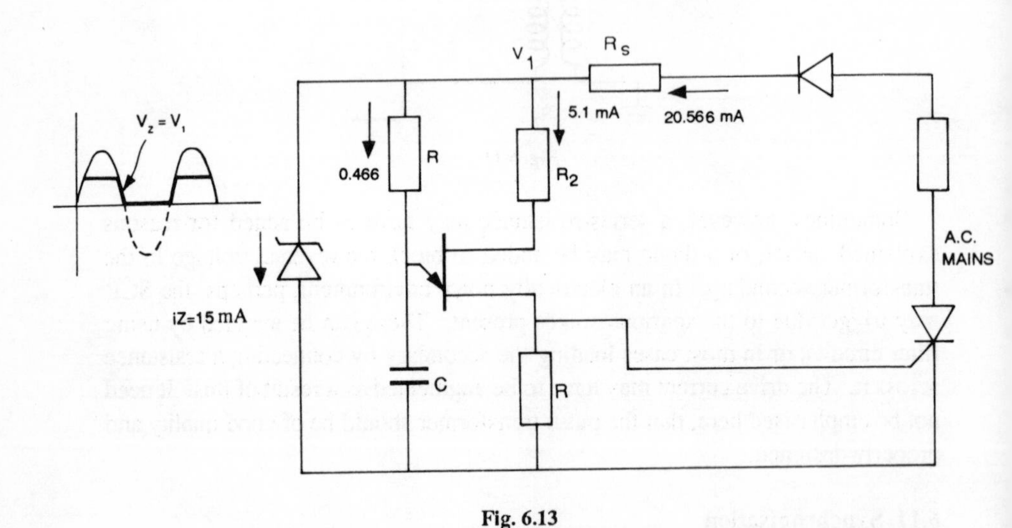

Fig. 6.13

Here, the supply voltage is derived from the same *ac* mains from which the SCR works. AC supply is rectified, and this rectified voltage is fed to a zener in series with a resistance. The voltage available across the zener is used as a supply voltage V_1, for the trigger circuit.

Since the voltage across the zener is pulsating voltage, becoming zero every time *ac* passes though zero, effectively the supply voltage, V_1, is withdrawn and is reapplied again as the second cycle begins. This ensures that the SCR gets a pulse in every cycle at precisely the same angle.

EXAMPLE 6.1

Design a half-wave controlled rectifier to give *dc* voltage of 50V to a resistive load of 50 Ohms from a 230V, 50Hz mains.

Solution

Refer Fig 6.13

For Half-wave controlled rectifier, we have,

$$V_{dc} = \frac{V_p}{2\pi}(1 + \cos\theta)$$

$$50 = \frac{230 \times \sqrt{2}}{2\pi}(1 + \cos\theta)$$

From which, $\theta = 91.95°$.

Hence, $t = \frac{91.95° \times 20 \text{ mS}}{360°}$, for 50Hz supply.

$$t = 5.1 \text{ ms.}$$

Assume $C = 0.1$ uF, and $V_{BB} = 24$V, and from the data-sheet for the UJT, we have

$$\eta_{min} = 0.56, \qquad \eta_{max} = 0.75$$

$$t = R_{max} \times C \ln \cdot \{1/(1 - \eta_{min})\}$$

i.e., $5.1 \times 10^{-3} = R_{max} \times 0.1 \times 10^{-6} \times \ln\{1/(1 - 0.56)\}$

from which, $R_{max} = 62.12$ k ohms.

Similarly,

$$5.1 \times 10^{-3} = R_{min} \times 0.1 \times 10^{-6} \times \ln\{1 + (1 - 0.75)\}$$

giving, $R_{min} = 36.78$ k ohms.

Let us check whether these values of the resistances will be practicable.
For UJT to fire,

$$R_{max} < \frac{V_{BB}(1 - \eta_{max})}{I_p}$$

where I_p is the peak current = 5uA, from the data sheets.

$$< \frac{24\,(1 - 0.75)}{5 \times 10^{-6}}$$

Therefore, $R_{max} < 1200$ k ohms.

For UJT to oscillate,

$$R_{min} > \frac{V_{BB} - V_v}{I_v}$$

$$R_{min} > \frac{24 - 1.8}{4 \times 10^{-3}}$$

where I_v is the valley current and = 4mA, from the data sheets. Therefore, $R_{min} >$ 5.55 k ohms.

Both the above conditions can be satisfied by using a combination of

$$R_f = 39 \text{ k ohms (fixed) and}$$
$$R_v = 22 \text{ k ohms (variable).}$$

$$\text{Maximum UJT current } I_{UJT(max)} = \frac{V_{BB}}{R_{BB(min)}}$$

$$= 24/4.7\text{k}$$
$$= 5.1 \text{ mA}$$

$$\text{Maximum current through } R = \frac{V_{BB} - V_v}{R_{min}}$$

$$= \frac{24 - 1.8}{39\text{k}}$$

$$I_{R(max)} = 0.466 \text{ mA}$$

We shall assume that the zener minimum current is 15mA. Therefore, maximum current through R_s,

$$I_{Rs} = 15 + 5.1 + 0.466$$
$$= 20.566 \text{ mA}$$
$$V_{Rs} = V_{supply}(\text{dc}) - V_z$$

$$= \frac{230 \times \sqrt{2} - 24}{3.14}$$

$$= 79.53\text{V}.$$

$$R_s = \frac{V_{Rs}}{I_{Rs}}$$

$$= 79.53\text{V} / 20.566\text{mA}$$

$$R_s = 3.86 \text{ k ohms}$$

Power rating of $R_s = (79.53)^2 / 3.86 \times 10^3$

$$= 1.64 \text{ W}.$$

We select R_s = 3.9 K-Ohms, 3 W.

SCR rating

$$PIV = 230 \times \sqrt{2}$$

$$= 325.27\text{V}$$

$$I_{RL} = 50/50$$

$$= 1 \text{ A}$$

Current rating of SCR should be at least 50% more than this value = 1.5 A

We select SCR as having a rating of 2A, 400 PIV

Pulse transformer

A 1:1 pulse transformer is used.

Power rating of R_f and R_v = 1/4W

We select the diode used in the Half-wave controlled rectifier as IN4004, 1A/400 PIV.

EXAMPLE 6.2

Design a full-wave controlled rectifier employing 2 SCRs and 2 diodes in bridge configuration to rectify 230V, 50Hz *ac* mains and give an output of 150V to a resistive load of 10 ohms. Use UJT (2N2646) trigger circuit with temperature compensation for triggering the SCRs. The minimum supply voltages for UJT trigger circuit, without temperature compensating resistor, are given for different values of condensors for proper triggering of SCRs to be used, in the following table

C *uf*	:	0.02	0.05	0.1	0.2	0.5	0.7	1.0
V_{BB} volts	:	34.0	24.0	16.0	11.5	11.0	10.5	10.0

Solution

For a full-wave rectifier,

$$V_{dc} = \frac{V_p}{\pi}(1 + \cos\theta)$$

$$150 = \frac{230 \times \sqrt{2}}{\pi}(1 + \cos\theta)$$

$$\therefore\ \theta = 63.33°$$

Now, T = 20ms for 360°

$$\therefore\ t = \frac{63.33°}{360°} \times 20$$

$$t = 3.52 \text{ ms}$$

UJT 2N2646 is used for triggering the SCR.
We select C = 0.1 microfarads, and from the data given in
C V/S V_{BB} table, V_{BB} = 16V.
Resistance R_2 is provided in the circuit for temperature compensation.

$$R_2 = \frac{10^4}{\eta\ V_{BB(\text{min})}}$$

$$= \frac{10^4}{0.66 \times 16} \quad \text{where } \eta = 0.66 \text{ (typical)}$$

$$= 946.96 \text{ ohms}$$

We select R_2 = 1 k ohms.
Now,

$$V_{BB}\,(\text{actual}) = \frac{(2200 + R_2)}{(2300)} \times V_{BB(\text{min})}$$

$$= \frac{(2200 + 1000)}{(2300)} \times 16$$

$$= 22.26\text{V}.$$

We select V_{BB} = 24V.
Recalculate R_2

We get $$R_2 = \frac{10^4}{\eta\ V_{BB}}$$

$$= \frac{10^4}{0.66\,(24)}$$

$$= 631.31 \text{ ohms.}$$

We select $R_2 = 680$ ohms.

Again,

$$V_{BB(\text{actual})} = \frac{(2200 + R_2)}{(2300)} V_{BB(\text{min})}$$

$$= \frac{(2200 + 680)}{(2300)} \times 16$$

$$= 20.03\text{V}$$

Therefore, $V_{BB} = 24$V is sufficient.

We select zener of 24V.

Now,

$$\eta_{min} = 0.56, \qquad \eta_{max} = 0.75$$

$$t = R\,C \ln \{1/(1-\eta)\}$$

$$3.52 \times 10^{-3} = R_{max} \times 0.1 \times 10^{-6} \ln \{1/(1-0.56)\}$$

$$R_{max} = 42.87 \text{ k ohms}$$

Similarly

$$3.52 \times 10^{-3} = R_{min} \times 0.1 \times 10^{-6} \ln \{1/(1-0.75)\}$$

$$R_{min} = 25.39 \text{ k ohms.}$$

For UJT to fire

$$R_{max} < \frac{V_{BB}\,(1-\eta_{max})}{I_{p(max)}}$$

$$< \frac{24\,(1-0.75)}{5 \times 10^{-6}}$$

$$< 1.2 \text{ M ohms.}$$

For UJT to oscillate,

$$R_{min} < \frac{V_{BB} - V_v}{I_v}$$

We assume $V_v = 1.8\text{V}$

$$R_{min} > (24 - 1.8)/4 \times 10^{-3}$$

$$> 5.5 \text{ k ohms}$$

Both the above conditions can be satisfied by selecting R as a combination of

R_f = 27 k ohms (fixed)
R_v = 22 k ohms (variable)

Maximum UJT current

$$I_{UJT(max)} = \frac{V_{BB}}{R_2 + R_{BB}}$$

$$= \frac{24}{680 + 4700}$$

$$= 4.49 \text{ mA}$$

Maximum current through R,

$$I_{R(max)} = \frac{V_{BB} - V_v}{R_{min}}$$

$$= \frac{24 - 1.8}{27 \times 10^3}$$

$$= 0.82 \text{ mA}$$

Let the minimum current through zener = 15 mA
Therefore, I_{max} through R_s = 15 + 4.5 + 0.82
= 20.32 mA.

EXAMPLE 6.3

(a) Design an *ac* power control circuit for firing angle of 90° to supply *ac* voltage to a resistive load of 10 ohms from 230V, 50Hz mains supply. Use back-to-back (inverse parallel) connected SCRs triggered by UJT circuit. No temperature compensating resistor is required in the trigger circuit. Use UJT type 2N2646. Specify the voltage and *dc* or average current ratings of the rectifier devices used and give specifications of the rest of the components. Calculate the *ac* rms voltage across the load resistance.

(b) Determine whether the SCRs used will definitely be triggered by the trigger circuit if the SCR used requires minimum supply voltage for UJT trigger circuit without temperature compensating resistor for different values of capacitance given by the following table

C *uf*	0.03	0.05	0.07	0.1	0.2	0.3	0.5	1.0	10
V_{BBmin}	34.0	26.0	23.0	20	16	15	13	12	11

Solution

Given

$$\theta = 90°$$

$$V_{dc} = \frac{V_m}{\pi}(1 + \cos\theta)$$

$$= \frac{230 \times /\sqrt{2}}{\pi}(1 + \cos 90°)$$

$$V_{dc} = 103.53 \text{ volts}$$

$$t = \frac{90°}{360°} \times 20 \text{ ms}$$

$$t = 5 \text{ ms}$$

We assume $C = 0.1$ microfarads; therefore, $V_{BB(min)} = 20V$
We select $V_{BB} = 24$ V, and a zener diode of 24V.
From the data sheet, we have $\eta_{min} = 0.56$, $n_{max} = 0.75$

$$t = R_{max} C \ln\left(\frac{1}{1-\eta_{min}}\right)$$

$$5 \times 10^{-3} = R_{max} \times 0.1 \times 10^{-6} \ln\{1/(1-0.56)\}$$

$$R_{max} = 60.9 \text{ k ohms}$$

Similarly,

$$5 \times 10^{-3} = R_{max} \times 0.1 \times 10^{-6} \ln\{1/(1-0.75)\}$$

$$= 36.06 \text{ k ohms}$$

For UJT to fire,

$$R_{max} < \frac{V_{BB}\,(1-\eta_{max})}{I_P}$$

$$< \frac{24\,(1-0.75)}{5\times10^{-6}}$$

$$< 1200 \text{ K-Ohms}$$

For UJT to oscillate,

$$R_{min} > \frac{V_{BB}-V_v}{I_v}$$

$$> \frac{24-1.8}{4\times10^{-3}}$$

$$> 5.55 \text{ k ohms}$$

The above conditions can be met by selecting R as a combination of

R_f = 33 k ohms (fixed)
R_v = 27 k ohms (variable)

Maximum UJT current, $I_{UJTmax} = V_{BB} / R_{BB}$
$= 24 / 4.7$
$= 5.1$ mA

Maximum current through $R = V_{BB} / R_{min}$
$= 24 / 33$
$= 0.72$ mA

Minimum current through zener diode I_z = 15 mA

Current through $R_s = 15 + 5.1 + 0.72$
$= 20.82$ mA

$V_{Rs} = V_{supply}(_{dc}) - V_z$
$= 2V_p - V_z$
$= 2 \times (230 \times /2) - V_z$
$= 183.07$V

$$R_s = \frac{V_{Rs}}{I_{Rs}}$$

$= 183.07 / 20.82$ mA
$= 8.79$ k-Ohms

We select $R_s = 8.2$ k ohms

Power rating of $R_s = (183.07)^2 / (8.2 \times 10^3)$

$= 4.08$ W

Hence, $R_s = 8.2$ k-Ohms and of 10W rating.

Power rating of R_f and $R_v = 1/4$ W.

Power rating of zener diode = 24 × 15 = 360 mW

We select a zener diode of 24V, having $P_{d(max)} = 0.5$W.

The 4 diodes for the bridge rectifier circuit should be of a comparable current rating. We select IN4001 diodes having current rating of 1A and PIV of 400V.

SCR ratings:

$$\text{PIV} = 230 \times \sqrt{2}$$

$$= 325.27 \text{ volts}$$

$$I_{RL(average)} = 103.53 / 10$$

$$= 10.353 \text{ A}$$

Current through each SCR = half the total average current

$= 5.1765$ A

Current rating should be at least 50% greater than the average current = 1.5 × 5.1765 = 7.76 A

We, therefore, select 2 SCRs of 10 A / 400 PIV each.

7

AMPLIFIERS

We shall discuss the various configurations of amplifiers in this chapter. There are mainly three main configurations of the amplifier normally employed in actual practice, namely the Common-base, the Common-collector and the Common-emitter. In this chapter we shall refer to the Common-emitter configuration.

7.1 Transistor Characteristics

Refer to Fig 7.1.

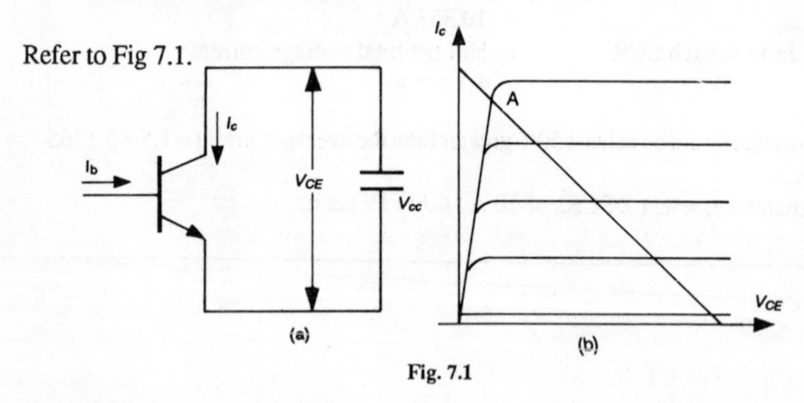

Fig. 7.1

The collector current, I_C, is governed by, and is a function of, two variables, namely, V_{CE} and I_B, the collector-to-emitter voltage and the base-current, respectively. Figure 7.1(b) shows the variation of the collector current with respect to V_{CE}.

The ratio I_C/I_B is called β (7.1)

and is the current-gain of the transistor. Usually, the collector-current and the emitter-current are assumed to be equal, since their difference is equal to the base-current, which is very small and can be considered negligible.

$$I_E = I_C + I_B \tag{7.2}$$

Since $I_C/I_B = \beta$, we therefore have

$$I_E = \beta \cdot I_B + I_B \tag{7.3}$$

Also

$$I_C = \alpha \cdot I_E \tag{7.4}$$

Substituting the values in equation 7.2, we get

$$\frac{I_C}{\alpha} = I_C + \frac{I_C}{\beta}$$

Therefore, $$1/\alpha = 1 + 1/\beta$$

Giving $$\alpha = \frac{\beta}{1+\beta} \tag{7.5}$$

Rewriting equation (7.5), we get

$$\beta = \frac{\alpha}{1-\alpha} \tag{7.6}$$

7.2 Common - Emitter Configuration

Let us discuss the common-emitter configuration. Refer to Fig 7.2

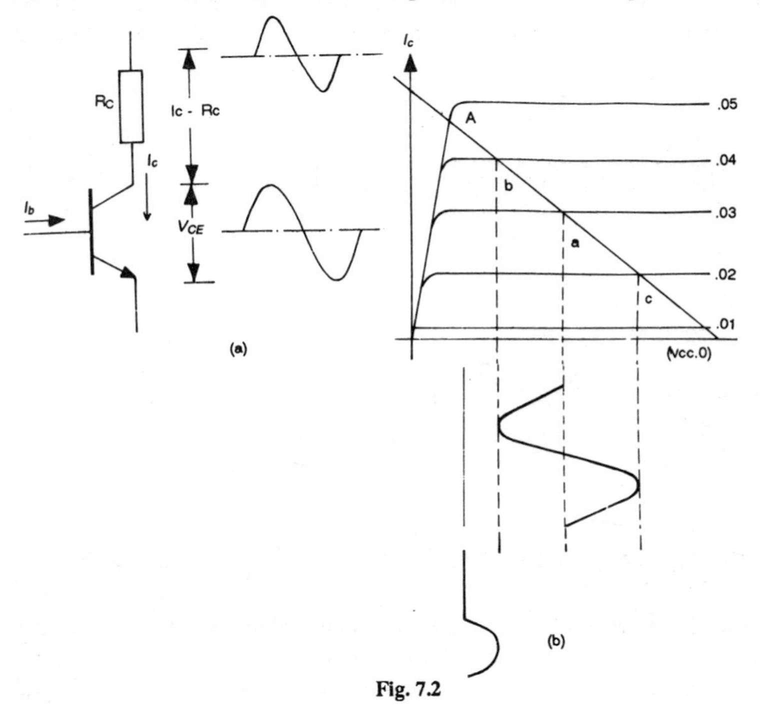

Fig. 7.2

From the figure it can be seen that,

$$V_{CC} = I_C \cdot R_c + V_{CE} \tag{7.7}$$

From this equation we can deduce that if we keep the voltage, V_{CC} constant, then the collector current will depend upon the base-current, I_B. The resistance R_c is also constant. Therefore, the only variables are the collector-to-emitter voltage and the collector-current. Obviously, if current I_C is increased, then the voltage

V_{CE} will have to decrease. If we consider the extreme values, then the voltage V_{CE} $= V_{CC}$ when the collector current is zero, and I_C will be maximum when the voltage V_{CE} is zero. This value of the current is equal to $\frac{V_{CC}}{R_c}$.

These two points are indicated in Fig. 7.2(b).

If these two points are joined by a straight line, we get the slope of the line which represents the resistance, R_c. It can be readily seen that the slope of the line will change only if the resistance R_c is changed. This line is called the **load-line.** It is very important to note that, in practice, voltage V_{CE} will never be equal to zero. The area covered by the linearly increasing part of the characteristics and the load-line near the current axis is called the saturation region.

In the set of characteristics given in Fig 7.2, it can be seen that if the base-current is 0.03mA, the point of operation is located where the $V_{CE} - I_C$ characteristic intersects the load-line, i.e., at a point *a*. If we increase the base-current to say, 0.04mA, the point of operation will move to point *b*. Likewise, if the point of operation is to be moved to point *c*, then the base-current will have to be reduced to 0.02mA. In general, the point of operation will always remain on the load-line.

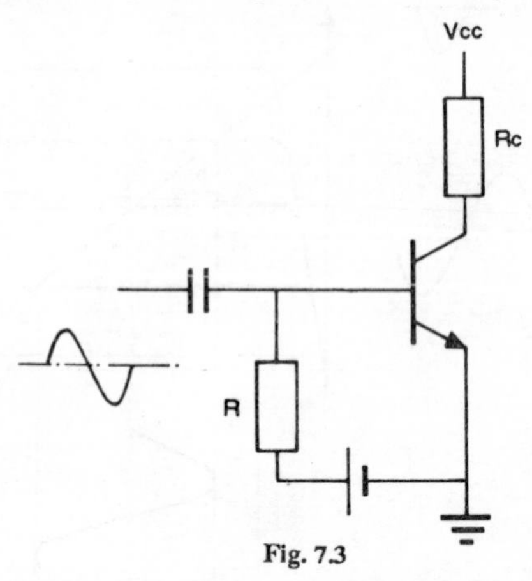

Fig. 7.3

Assume that a base-current equal to, say, 0.03mA is being forced in the transistor by some means, in our case by a voltage V through a resistance *R*, as shown in Fig 7.3.

A sinusoidal voltage is also applied to the circuit which will force, and super impose, a sinusoidal current on the *dc* base-current. Further, we assume that the magnitude of this sinusoidal voltage is such that it is able to increase the base-current from 0.03mA to 0.04mA, when the signal is at positive maximum, and reduce to 0.02mA when the signal is at negative maximum. Thus, the signal is able to change the current sinusoidally by ±0.01mA from its initial value of 0.03mA. The point of operation will move from **a** to **b** and back to **a** during the positive half-cycle of the signal, and from **a** to **c** and back to **a** during the negative half-cycle. This change in the base-current will force a change in the collector-current which will be β times larger and, therefore, the voltage V_{CE} will also change as per Equation 7.7.

Properly designed, this change in V_{CE} will be sinusoidal, and almost an exact replica of the signal fed at the base-circuit, but its magnitude will be much larger. The circuit is, then, said to be working as an amplifier. As can be seen from the diagram, the output voltage is 180° out-of-phase to the signal wave-form.

7.3 Biasing

It should be clear from the above discussion that the output voltage will be exact reproduction of the input signal voltage only if the input voltage is able to force the change in the base-current *linearly*. Take a case where the initial point of operation, the point a, also called the **quiscent point**, is at A (refer to Fig 7.2). As the base-current increases as a result of the positive transition of the signal voltage, the quiscent point at A is not able to move up the load-line, since it is a point where the saturation region begins. That means that the output voltage is minimum and can not decrease further. The output voltage, therefore, can not be an amplified reproduction of the input voltage, at least during the positive half-cycle of the signal voltage. The negative half-wave, however, will be amplified since the change in the base-current will be able to move the operating point linearly. Similarly, when the quiscent point is at a point **B**, the negative half-wave of the signal voltage will not be reproduced in the output circuit but a positive half-wave will be able to move the quiscent point up the load line linearly and hence, will be reproduced in the output circuit.

This leads us to an important decision: that the point of operation, the quiscent point, should be kept at the center of the load-line, or somewhere on the load-line such that the signal transition will be able to move it up and down the load-line with simple harmonic motion, provided that at no time it is forced to go beyond the point A or below the point *b*. These two points, the point A and the point *b*, are, respectively, called the saturation point and the cut-off point. The initial base-current, which decides the position of the quiscent point, is called the **biasing current.** We shall now discuss, briefly, the methods of biasing and their stability.

7.4 Stability

There are two factors which affect the selection of the Q-point. Assume that the biasing current is so selected that the Q-point is placed at a point Q_1. Further assume that the transistor used in the circuit is replaced by another transistor. Even if the second transistor belongs to the same batch of manufacture, its current gain, among other parameters, may be different than that of the first transistor. So much so, that the current gain of a transistor is specified between minimum and maximum values. For the same variation of the base-current, the collector-current and, hence, the voltage V_{CE} variations could be different for both the transistors. The new Q-point, Q_2, is shown in the figure. It is assumed that the second transistor is having a greater current gain than the first, producing greater spacing between the characteristics.

1. Refer to Fig 7.4.

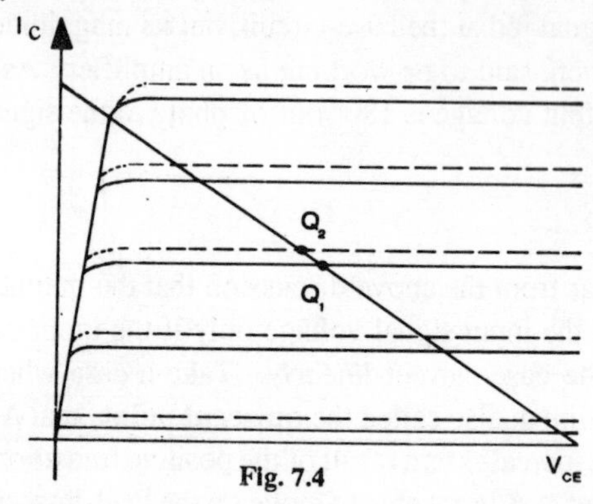

Fig. 7.4

For the same magnitude of the signal voltage, the transition provided to the second Q-point could force the transistor to the saturation region. This shows that to provide the stability against device variation, i.e., against the change in the current-gains, the Q-point base-current should be allowed to change.

2. More severe is the instability caused by the temperature variations. The reverse saturation current, I_{CBO}, virtually doubles with every 10°C rise in temperature. This gives rise to a regenerative cycle. The collector current, I_C, gives rise to power dissipation, which increases the temperature. This increase in the temperature increases the reverse current, which in-turn, increases the collector current to that extent, further increasing the temperature. This regenerative cycle, in unfortunate circumstances, may burn-out the device, or at least, certainly force the Q-point to shift.

It can be shown that the collector current for a common-emitter amplifier is given by the equation

$$I_C = \beta \cdot I_B + (\beta + 1) \cdot I_{Co} \tag{7.8}$$

The current I_{Co} here is written in place of the current I_{CBo}. The collector current I_C, as can be seen in the above equation, is dependent on three variables, all of which are, to varying degrees dependent upon temperature.

The current gain, β, is very strongly dependent on temperature. The base-current, I_B, is also dependent on temperature. This current increases with temperature, indicating that the input resistance decreases with temperature. The net effect is the same as if the voltage, V_{BE}, changes at the rate of about – 2.5mV/°C. The third variable I_{CBo}, changes to the extent of doubling for every 10°C rise in temperature.

As has already been mentioned, the temperature rise will force the Q-point to shift. This is also evident from Equation 7.8. As the temperature increases, all the three variables will show upward trend, increasing the collector-current I_C. This will force the curves of the characteristics upwards, causing for the same magnitude of signal, distortion in the output-voltage, or saturation in the worst case. What could have been an intelligent selection of a Q-point at one operating temperature, could cause severe problems at some different temperature.

Again referring to Equation 7.8, it is noticed that, if I_C increases due to increase in the temperature or device replacement, the initial value can be restored if we reduce the base-current. Likewise, as already pointed out, the increase in the collector-current due to increased I_{Co}, because of increase in the temperature, is a regenerative cycle. This can also be prevented by forcing I_B to decrease.

Stability factor is a measure of the extent to which the collector-current can be kept constant in spite of the variation in the current I_{Co} because of the temperature. Assuming that both, the current I_B and the current gain, β, are constant, we have,

$$S = \frac{dI_C}{dI_{Co}} \tag{7.9}$$

If we differentiate Equation 7.8 with reference to I_C, we get,

$$I = \beta \cdot \frac{dI_B}{dI_C} + \cdot \frac{1+\beta}{S}$$

or,

$$S = \frac{1+\beta}{1-\beta\,(dI_B/dI_C)} \tag{7.10}$$

The S in the above equation is called the **stabilisation factor**, and from Equation 7.9, it is clear that this factor should be as small as possible, since we want that the change in the collector-current with reference to unit change in I_{Co}, the leakage current, to be as small as possible, ideally, zero.

With this introduction to the stability factor, let us see some of the more common biasing circuits.

7.5 The Fixed-Bias Circuit

Refer to Fig 7.5

The *dc* analysis of the circuit yields the following equation

$$V_{CC} = (I_B \times R_B) + V_{BE}$$

or,

$$I_B = (V_{CC} - V_{BE})/R_B \tag{7.11}$$

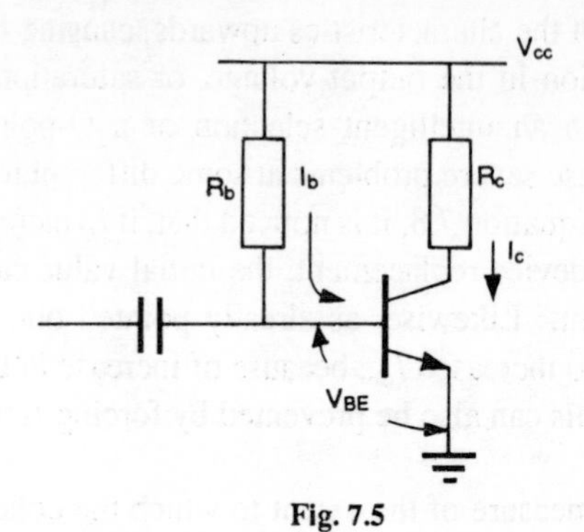

Fig. 7.5

Since the equation is independent of the current I_C, the stability Equation 7.10, reduces to

$$S = 1 + \beta \tag{7.12}$$

This method of biasing is frequently called base-current biasing, and as can be seen if the current gain of transistor is, say, 100, then the stability factor = 101, which is a very poor figure indeed.

The advantages of this method are:

1. Simplicity
2. Small number of components required, and
3. If the supply voltage is very large as compared to V_{BE} of the transistor, then from Equation 7.11, the base current becomes largely independent of the voltage V_{BE}.

The disadvantage, evidently, is that the leakage current, I_{Co}, is not controlled and can cause very serious problems as the temperature increases.

7.6 Collector-to-Base Biasing

Refer to Fig 7.6.

This circuit is the simplest way to provide some degree of stabilisation to the amplifier operating point. The loop equation for this circuit can be written as:

$$V_{CC} = (I_B + I_C) \cdot R_c + I_B \cdot R_B + V_{BE} \tag{7.13}$$

i.e.,

$$I_B = \frac{V_{CC} - V_{BE} - I_C \cdot R_c}{R_c + R_B} \tag{7.14}$$

Hence,

$$\frac{dI_B}{dI_C} = -\frac{R_c}{R_c + R_B},$$

From Equation 7.10, we have

$$S = \frac{1+\beta}{1+\beta\left(\frac{R_c}{R_c+R_b}\right)} \tag{7.15}$$

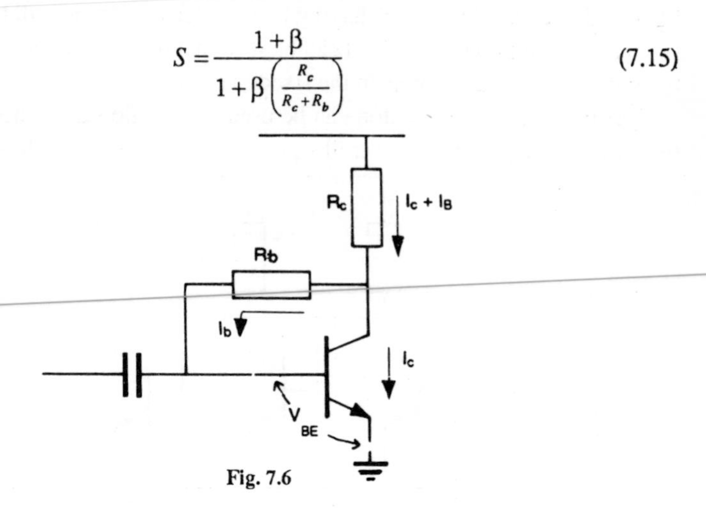

Fig. 7.6

As can be seen, this value of the stability factor is smaller than the value obtained by the fixed bias circuit.

The main disadvantage of the circuit, apart from providing a fairly small amount of stabilisation, is that it provides gain degeneration by providing a negative feedback. If the signal voltage causes increase in the base-current, it will increase the collector current. This increase in the collector-current will force V_{CE} to decrease, in-turn, decreasing the base current, as per Equation 7.14. Thus, the net increase in the base-current due to signal voltage is smaller with negative feedback than in the circuit where R_B is connected to a fixed voltage for the bias. However, this kind of signal-gain degeneration can be reduced to a great extent by splitting the resistance R_B in two parts, and connecting a condenser from the junction point of these two resistors to the ground, as shown in Fig 7.7. The condenser should provide a low impedance path at the signal frequency, effectively forcing the fed back signal to bypass the base circuit.

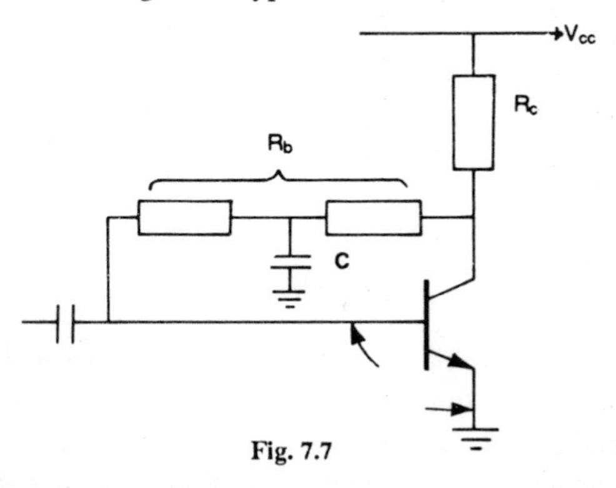

Fig. 7.7

It should be noted that this type of stabilisation is effective only for relatively light load circuits. Amplifiers having large collector-current will have, obviously, small R_c. From Equation 7.15, it can be seen that if the resistance R_c is small, then the stabilisation is as poor as in the fixed bias method.

This method of stabilisation can be used in a single stage amplifier, or a multistage amplifier where the overall signal inversion occurs as shown in Fig 7.8.

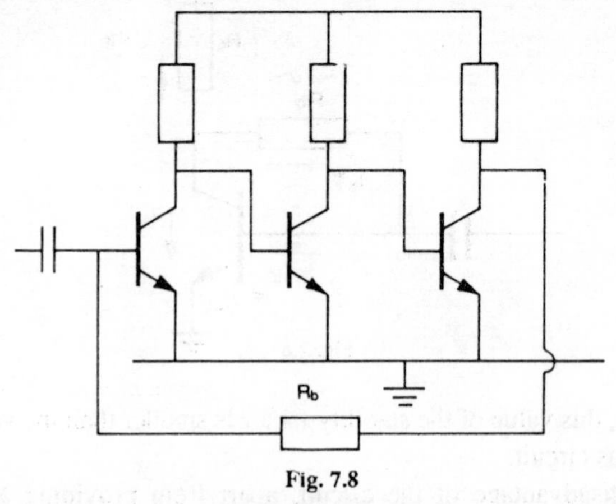

Fig. 7.8

7.7 Potential Divider Type or Self Bias

A biasing circuit that can be used even for low collector resistances, is shown in Fig 7.9.

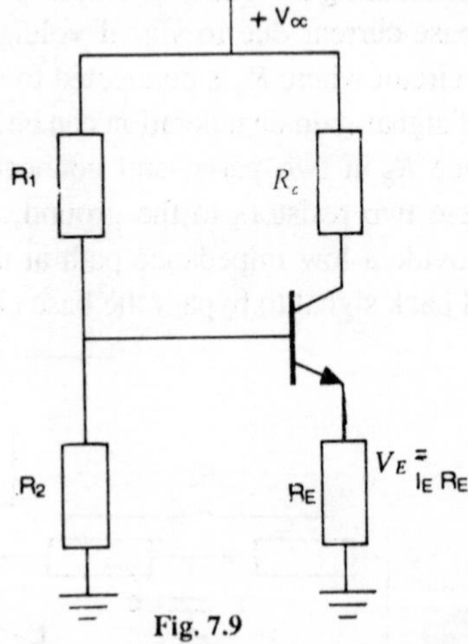

Fig. 7.9

The resistance R_E develops a voltage across it equal to $I_E \times R_E$. Since the base-emitter junction has to be forward biased, a potential divider network of the resistances R_1 and R_2 is used to provide the required voltage at the base. Figure 7.10 shows Thevenin's equivalent circuit for finding the base-current.

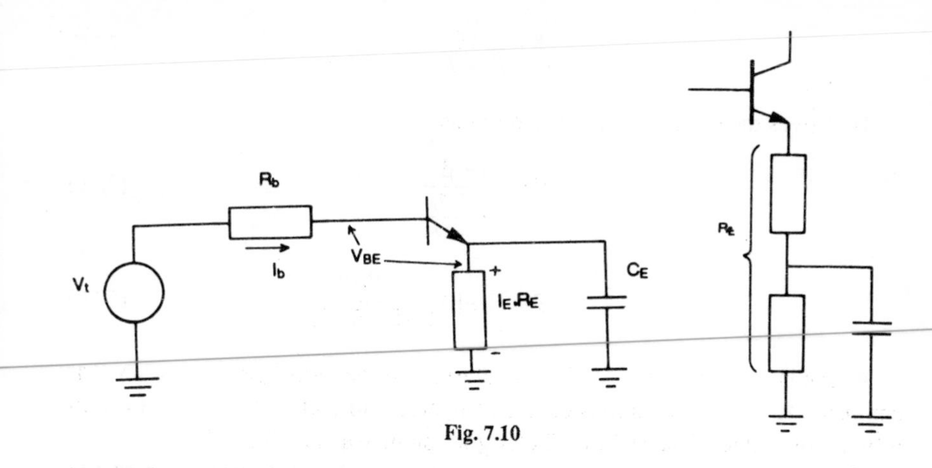

Fig. 7.10

It should be noted here that as per Equation 7.8, if the value of the resistance R_B, is kept small, the value of the current I_B will become sufficiently large so that the variations in the current, I_{Co}, do not greatly matter. In other words, the voltage at base becomes independent of the leakage current.

The resistance R_b, in our case, is a Thevenin's equivalent resistance, and is equal to the parallel combination of resistances R_1 and R_2.

It should also be noted that if the temperature forces the base current to increase, in turn the emitter current will also increase. This increase in the emitter current will develop a larger voltage across the resistance R_E. Since this voltage opposes the voltage at the base in the base current loop, increase in the voltage at the emitter would mean a reduction in the base current. This effect is the negative feed-back effect, similar to that achieved in the collector-to-base biasing circuit. Thus, the increase in the base current, and hence, in the collector current, is less than would have been the case if resistance, R_E, were absent.

Applying Thevenin's theorem to the circuit of Fig 7.9, for finding the base current, we have,

$$V_t = \frac{R_2 \times V_{cc}}{R_1 + R_2}$$

and

$$R_b = \frac{R_1 R_2}{R_1 + R_2} \tag{7.16}$$

Thevenin's equivalent around the base, is shown in Fig 7.10. The loop equation around the base circuit can be written as:

$$V_t = I_B \cdot R_b + V_{BE} + (I_B + I_C) \times R_E \tag{7.17}$$

Differentiating this equation with respect to I_C, we get,

$$\frac{dI_B}{dI_C} = \frac{R_E}{R_E + R_b} \tag{7.18}$$

This gives the equation for the stability as,

$$S = \frac{1+\beta}{1+\frac{\beta \cdot R_E}{R_E + R_b}} \tag{7.19}$$

$$S = (1+\beta)\frac{1+R_b/R_E}{1+\beta+R_b/R_E} \tag{7.20}$$

As can be seen, the value of S is equal to one if the ratio R_b/R_E is very small as compared to 1. As this ratio becomes comparable to unity, and beyond towards infinity, the value of the stability factor goes on increasing till $S = 1 + \beta$.

This improvement in the stability upto a factor = 1 is achieved at the cost of power dissipation. To improve the stability, the equivalent resistance R_b should be decreased, forcing more current in the voltage divider network of R_1 and R_2.

Often, to prevent the loss of gain due to the negative feedback, the resistance R_E is either partially or fully shunted by a capacitor C_E, as shown in the diagram. The value of the capacitance should be such as to offer reactance X_{CE} equal to about one-tenth of the value of the resistance R_E, at the lowest operating frequency.

7.8 Cascaded Stages

If several CE amplifier stages are connected in cascade for the purpose of increasing the overall gain, there are several aspects that have to be considered. Such an amplifier system suffers from finite bandwidth. Also, normally, the output of one stage is connected to the input of the succeeding stage through a coupling capacitor, C_c.

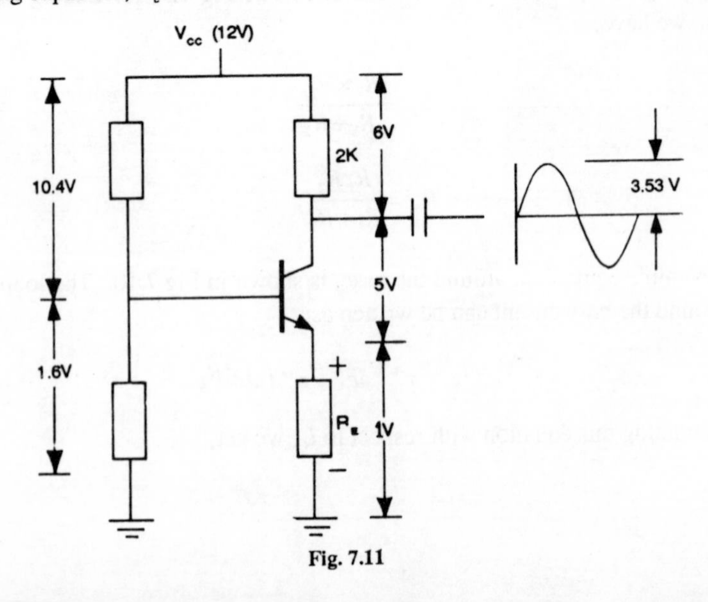

Fig. 7.11

Figure 7.11 indicates such a cascade. In a long chain of cascaded stages, the input impedance, Z_i, of each stage is identical.

i.e., $$Z_{i1} = Z_{i2} = \ldots = Z_{in} \tag{7.21}$$

Also $Z_i = h_{ie}$, if the parallel combination of the resistances R_1 and R_2, i.e., R_b, is very large as compared to h_{ie}; if not, then,

$$Z_i = h_{ie} + R_b \tag{7.22}$$

The actual input impedance of the second and subsequent stages will be equal to the parallel combination of R_c of the previous stage and Z_i of the following stage. Likewise, if R_o is the output resistance of the stage, then, the load resistance, R_L, of the stage is given by the parallel combination of the load resistance of the stage, R_L, ($= R_o$) and Z_i of the following stage.

i.e., $$R'_{o1} = \frac{R_{c1} \times Z_{i2}}{R_{c1} + Z_{i2}} \tag{7.23}$$

where Z_{i2} can be made equal to h_{ie2}, as per Equation 7.22.

EXAMPLE 7.1

Design a single stage CE amplifier to give a voltage-gain of 100 or more, with the temperature stability factor, $S = 10$ and the output voltage 2.5V rms.
Use BC147B transistor and the V_{CC} supply of 12V.

Solution

Refer to Fig 7.12. (Numerical values in the figure refer to the next problem).

The specifications for the transistor BC147B are:

$$\begin{aligned}
\boldsymbol{P_{d(\max)}} &= \mathbf{250\ mW} \\
\boldsymbol{I_{c(\max)}} &= \mathbf{100\ ma} \\
\boldsymbol{h_{FE}} &= \mathbf{200\ (min),\ 290\ (typ),\ 400\ (max).} \\
\boldsymbol{h_{fe}} &= \mathbf{240\ (min),\ 320\ (typ),\ 500\ (max).} \\
\boldsymbol{h_{ie}} &= \mathbf{4.5\ K} \\
\boldsymbol{h_{oe}} &= \mathbf{30\ \mu\text{-}mhos.} \\
\boldsymbol{h_{re}} &= \mathbf{2 \times 10^{-4}.}
\end{aligned}$$

Calculation of the collector load resistance, R_c

Since, $$h = h_{ie} \cdot h_{oe} - h_{fe} \cdot h_{re}$$

= 0.071, taking the values of the h-parameters from the data sheets which are reproduced above, for a voltage gain of more than 100, we have,

$$A_v = \frac{h_{fe(\min)} \cdot R_L}{h_{ie} + \Delta h \cdot R_L}$$

The value of $h_{fe_{(\min)}}$ is selected as the gain needed is minimum 100.

Therefore, $$100 = \frac{240 \times R_L}{4.5\text{K} + 0.071 R_L}$$

from which $$R_L = 1932$$

We select a nearest higher value of the resistance since the gain required is minimum 100.

Hence, we select $R_L = 2\text{K}$.

Selection of the operating point

$$V_{o(\text{rms})} = 2.5\text{V}.$$

$$V_{o(\text{peak})} = \sqrt{2} \times 2.5$$

$$= 3.53 \text{ Volts.}$$

We shall select the operating point V_{CEQ}

$$= V_{o(\text{peak})} + \text{about } 1.5\text{V}.$$

$$= 3.53 + 1.5 = 5.03 \text{ Volts, say, } 5\text{V}.$$

1.5 Volts added should take care of the saturation voltage, and also avoid reaching the saturation region.

As a thumb-rule, we shall take V_{RE}, the voltage at the emitter of the transistor, and hence, voltage across the resistance R_E, as about 10 to 20% of the supply voltage, V_{CC}. Taking this as about 10%, we have $V_{RE} = 1\text{V}$.

Therefore, $$V_{Rc} = V_{CC} - (V_{CEQ} + V_{RE})$$

$$= 12 - (5 + 1)$$

$$= 6\text{V}$$

$$I_{CQ} = V_{RC} / R_c$$

$$= 6\text{V} / 2\text{K} = 3\text{mA}.$$

Calculation of the resistance R_E

Assuming the emitter current to be equal to the collector current, we have,

$$R_E = 1\text{V} / 3\text{mA}$$

$= 333.33$ ohms.

We select the resistance $R_E = 330$ ohms.

The power dissipation in the resistance $R_E = (3mA)^2 \times 330$ ohm

$= 2.97$ mw

We select resistance $R_E = 330$ ohms, 0.25W

Selection of the resistances R_1 and R_2.

Since the stability required is 10, from the equation of the stability, we have,

$$\frac{R_E}{R_E + R_B} = 0.0969$$

from which, $R_B = 3075.9$ ohms.

also $$R_B = \frac{R_1 \times R_2}{R_1 + R_2} = 3075.9 \text{ ohms} \qquad \text{(i)}$$

$$V_{R2} = V_{RE} + 0.6V$$
$$= 1.0 + 0.6$$
$$= 1.6V.$$
$$V_{R1} = V_{CC} - V_{R2}$$
$$= 12 - 1.6$$
$$= 10.4V.$$

Hence, $$(R_1 / R_2) = (10.4 / 1.6)$$
$$= 6.5 \qquad \text{(ii)}$$

From Equations (i) and (ii), we have

$$R_2 = 3459.1 \text{ ohms.}$$

and $$R_1 = 23069.2 \text{ ohms.}$$

We select the nearest available standard values, giving,

$$R_2 = 3.3\text{K, and}$$
$$R_1 = 22\text{K.}$$

EXAMPLE 7.2

Design a single stage voltage amplifier with $S = 9$, $A_v = 150$, the output voltage required is 2.5V rms.

Assume the supply voltage = 9.0V

Also design a coupling capacitor to be used if two such stages are to be coupled. The frequency range of the operation is to be between 20 Hz and 20KHz.

Calculate the voltage gain of each of the stages, and also the input and the output resistance for the two stage amplifier.

Solution

Refer to Fig 7.12

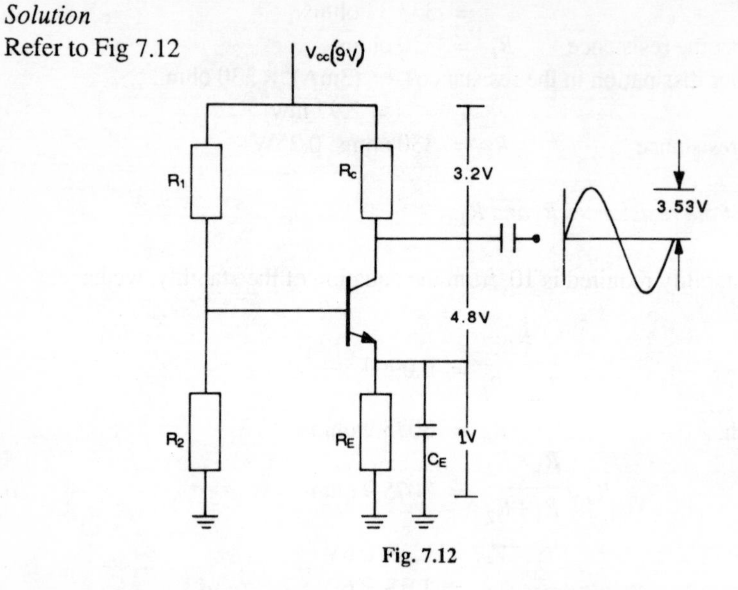

Fig. 7.12

If the supply voltage is not given, we may assume a supply voltage as about 4 times the output voltage required. Here, the supply voltage is given as 9V.

We again select a small signal, low-power transistor BC147B for the purpose. The data is given in Example 7.1.

Given, $V_{o(rms)} = 2.5\text{ V}$

$$V_{o(peak)} = \sqrt{2} \times 2.5$$

$$= 3.53\text{ V}$$

now, $V_{CEQ} > V_{CE(sat)} + V_{o(peak)}$

$$> 0.25 + 3.53$$

$$> 3.787$$

We select $V_{CEQ} = 4.8\text{ V}$, (about 1 V more.)

now, $V_{cc} = V_{RC} + V_{CEQ} + V_{RE}$

We assume $V_{RE} = 1\text{ V}$. (About 10% of V_{cc})

therefore, $V_{RC} = V_{cc} - V_{CEQ} - V_{RE}$

$$= 9 - 4.8 - 1$$

$$= 3.2\text{ V.}$$

Step III Selection of R_E

$$I_{CEQ} = V_{RC} / R_c$$

$$= 3.2 / 3.3\text{k}$$

$$= 0.969\text{ mA.}$$

Assuming $I_E = I_{CEQ}$

$$R_E = V_E / I_E$$

$$= 1 / 0.969$$

$$= 1.031 \text{ k ohms.}$$

We select $R_E = 1$ k ohm.

Power dissipated through the resistance,

$$R_E = (0.969 \times 10^{-3})^2 \times (1 \times 10^3)$$
$$= 0.938 \text{ mW.}$$

We select $R_E = 1$ k ohm, 0.25 W.

Step IV Selection of R_1 and R_2

Stability factor

$$S = \frac{(1+B)}{1+\frac{B R_E}{(R_E+R_B)}}$$

$$S = \frac{(1+B)}{(1+Bk)}, \text{ where } k = \frac{R_E}{R_B+R_E}$$

Since $S = 9$ is required, we choose $\beta = h_{FE(\text{typical})}$

$$= 290.$$

Therefore,

$$9 = \frac{1+290}{1+(290\times k)}$$

Hence

$$k = 0.108$$

$$0.108 = \frac{R_E}{R_B+R_E}$$

$$= 1/(R_B + 1)$$

$$0.108\, R_B = 0.892$$
$$R_B = 8.26 \text{ k ohms.}$$

$$R_B = \frac{R_1 \times R_2}{R_1+R_2}$$

$$= 8.26 \text{ k ohms.} \quad \text{(i)}$$

Also

$$V_{R2} = V_{RE} + 0.6$$
$$V_{R2} = 1.6 \text{ V}$$

and

$$V_{R2} = \frac{V_{CC} \times R_2}{R_1+R_2}$$

$$1.6 = \frac{9 \times R_2}{R_1 + R_2}$$

giving

$$\frac{R_2}{R_1 + R_2} = 0.177$$

$$R_2 = 0.177R_1 + 0.177R_2$$
$$0.822R_2 = 0.177R_1$$
$$R_1 = 4.645\,R_2 \quad \text{(ii)}$$

From equation (i) and (ii), we get

$$\frac{4.465 \times R_2 \times R_2}{5.645 \times R_2} = 8260$$

From which, R_2 = 10.04 k ohms.
We select R_2 = 10 k ohms.
Therefore, R_1 = 46.45 k ohms.
We select R_1 = 47 k ohms, 0.25W
R_2 = 10 k ohms, 0.25W

Step V Selection of C_E

Let X_{CE} at $20H_z$ = 1/10 of R_E
X_{CE} = 1/10 (1000)
= 100 ohms.

$$100 = \frac{1}{2 \times 3.14 \times 20 \times C_E}$$

$$C_E = 1/(2 \times 3.14 \times 20 \times 100)$$
C_E = 79.57 uF
We select C_E = 100 uF.

$$R_b = \frac{R_1 R_2}{R_1 + R_2}$$

= 8.245 k ohms

Input impedance of 1st stage

$Z_i = h_{ie}$
= 4.5 k ohms.

Therefore, $Z'_i = (R_b \times Z_i)/(R_b + Z_i)$

$$Z'_i = \frac{8.245 \times 4.5}{8.245 + 4.5}$$

$$= 2.911 \text{ k ohms.}$$

Since the input impedances of all the stages are assumed to be identical, the input impedance of the second stage = Z_{i2}

$$= 2.911 \text{ k ohms.}$$

Output load impedance of the 1st stage,

$$R_{01} = \frac{R_{c1} \times Z'_{i2}}{R_{c1} + Z'_{i2}}$$

$$= \frac{3.3 \times 2.911}{3.3 + 2.911}$$

$$= 1.546 \text{ k ohms.}$$

The voltage gain of the 1st stage, therefore, is equal to,

$$A_{v1} = \frac{-h_{fe} R'_L}{h_{ie}}$$

$$= \frac{-330 \times 1.546K}{4.5K}$$

$$= 113.37$$

Voltage gain of the IInd stage

$$A_{v2} = \frac{-h_{fe} R'_{L2}}{h_{ie}}$$

$$= \frac{-330 \times 3.3}{4.5}$$

$$= 242$$

Overall voltage gain $A_v = A_{v1} \times A_{v2}$

$$= 113.37 \times 242$$

$$= 27435.54$$

Output impedance $R_{out} = R_{c2} = 3.3$ k ohms.

Selection of coupling capacitor C_c

For frequency = 20Hz,

$C_c = 1/(2 \times 3.14 \times f(R_o + R_i)$

Where R_o = output impedance of 1st stage

R_i = input impedance of 2nd stage

$R_i = Z'_{i2} = 2.911$ k ohms.

$R_o = R'_{L1} = 1.546$ k ohms.

$C_c = 1/(2 \times 3.14 \times 20 \times (2.911 + 1.546) \times 10^3)$

$= 1.78$ uF

We select $C_c = 1.8$ uF.

EXAMPLE 7.3

Design a single stage transistor voltage amplifier with $A_v = 50$, $V_{cc} = 10$ V, and $R_c = 500$ ohms. If two identical stages as designed above are to be cascaded, calculate the value of the coupling capacitor so that the lower 3 *dB* frequency is 20Hz.

Solution

$A_v = h_{FE} R_c / h_{ie}$

$50 = h_{FE}/h_{ie} \times 500$

$h_{FE} / h_{ie} = 0.1$

We select a transistor with $h_{FE} / h_{ie} >= 0.1$

We assume that this transistor has

$h_{FE} = 100$ and $h_{ie} = 1$ k ohm.

$V_{cc} = V_{CEQ} + V_{RC} + V_{RE}$

Choose $V_{CEQ} = V_{cc}/2 = 5$ V

Let $V_{RE} = 2$ V

Therefore, $V_{RC} = 10 - 5 - 2$

$= 3$ V

$I_{CQ} = V_{RC} / R_C$

$= 3/500$

$= 6$ mA

$R_E = V_{RE} / I_{CEQ}$

$= 2/6$

$= 333$ ohms.

We select $R_E = 330$ ohms.

$R_2 = 0.1 \times h_{FE(min)} \times R_E$

$= 0.1 \times 50 \times 330$

$= 1650$ ohms.

We select $R_E = 1.6$ k ohms.

Also

$$I_b = I_{R1} - I_{R2} \text{ and}$$

$$V_B = V_{BE(on)} + V_{RE}$$

$$= 0.7 + 2 = 2.7 \text{ V}$$

$$\frac{I_c}{h_{FE(min)}} = \frac{V_{cc} - V_D}{R_1} - \frac{V_D}{R_2}$$

$$\frac{6 \text{ mA}}{50} = \frac{10 - 2.7 - 2.7}{R_1 \; 1.616}$$

From which $R_1 = 4.038$ k ohms.

We select $R_1 = 3.9$ k ohms.

Selection of C_E

Since X_{CE} at 20Hz $= 0.1 \times R_E$

$$X_{CE} = 33 \text{ ohms.}$$

$$C_E = 1 / (2 \times 3.14 \times 20 \times 33)$$

$$= 241.14 \text{ uF}$$

$\therefore$ We select $C_E = 250$ uF.

Selection of C_C

For lower 3 *dB* frequency = 20 Hz

$$C_C = 1 / (2 \times 3.14 \times f(R_{o1} + R_{i2})$$

Where R_{o1} = output impedance of 1st stage, and

R_{i2} = input impedance of 2nd stage

$$R' = \frac{R_1 \times R_2}{R_1 + R_2} = \frac{4 \times 1.6}{4 + 1.6} = 1.142 \text{ k ohms.}$$

$$R_o = \frac{R_c / h_{oe}}{R_c + (1/h_{oe})}$$

$$= 500 \text{ ohms.}$$

$$R_i = \frac{R' \times h_{ie}}{R' + h_{ie}}$$

$$= \frac{1.142 \times 1}{1.142 + 1}$$

$$= 533.3 \text{ ohms.}$$

$$C_C = 1 / (2 \times 3.14 \times 20 \, (500 + 533.33))$$

$$= 7.7 \text{ uF}$$

$\therefore$ We select $C_C = 10.0$ uF

8

JFET AMPLIFIERS

8.1 Overview of JFETs

The Field Effect Transistor is a semiconductor device whose operation depends on the electric field. The transistor has three terminals, namely, the source, the drain and the gate. For its operation refer to Figure 8.1.

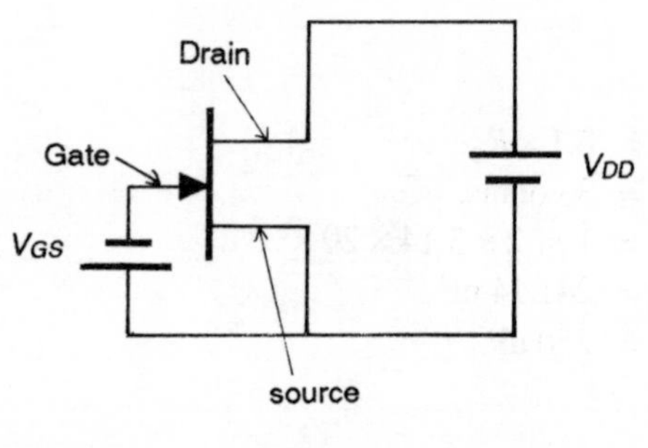

Fig. 8.1

In this figure, the symbolic representation is given for n-channel JFET. A voltage V_{DD} is applied to the drain positive and the negative end of the supply is connected to the source of the transistor. The gate is applied a negative voltage, V_{GS}, with reference to the source.

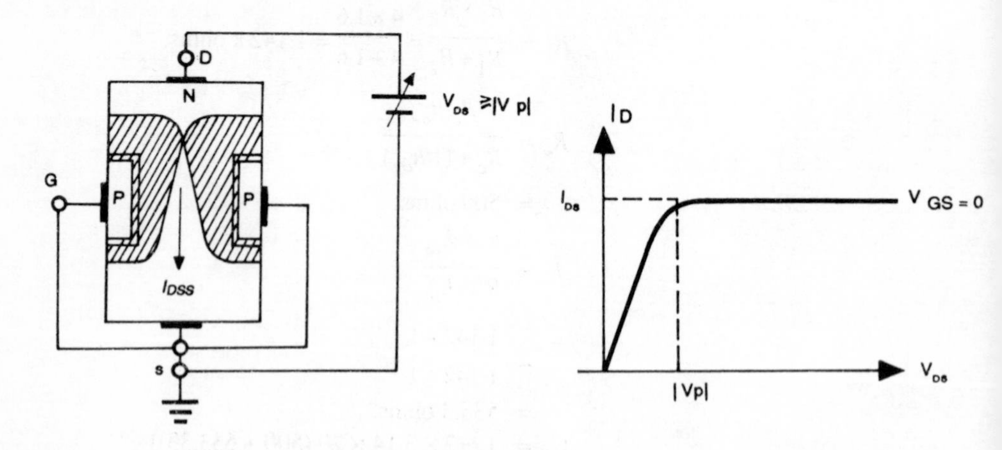

Fig. 8.2

Figure 8.2 shows the basic structure of the JFET, with the voltages applied as given in Fig 8.1. The drain and the source are the two end-terminal connections of the *n*-type semi-conductor, called a **channel**. The drain-to-source current, I_{DS}, is a function of two variables, V_{DS} and V_{GS}. Since the gate is *p*-type, forming a junction with the channel, which is an *n*-type semiconductor, giving a negative voltage to the gate with reference to the source produces a depletion region at the junction in the channel.

For a particular constant value of the gate voltage, say, – 2.0V, the dependence of I_{DS} on V_{DS} is as shown in Fig 8.3.

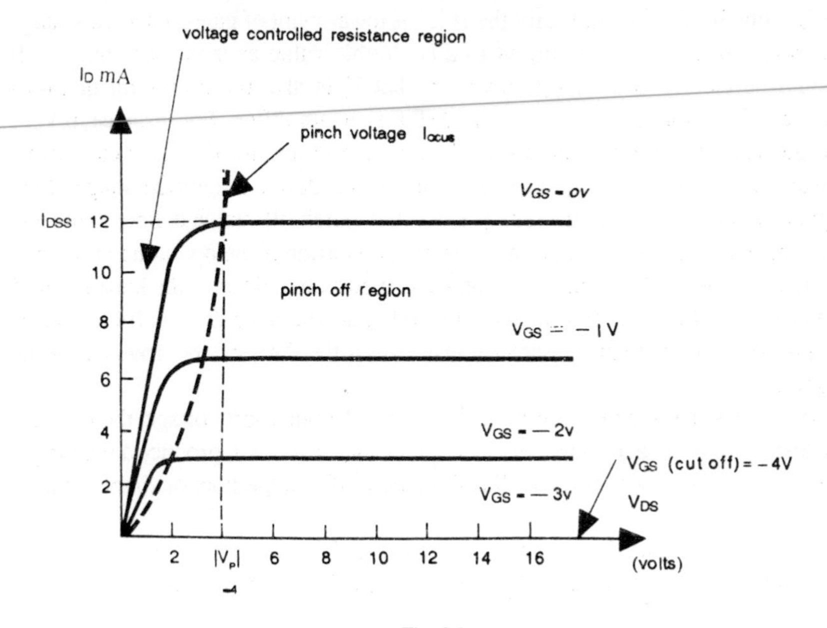

Fig. 8.3

Upto a certain value of V_{DS}, the current, I_{DS}, goes on increasing proportionally. This part of the characteristic follows the ohmic rule. After a certain value of the V_{DS}, the saturation of the current gradually takes place. The slope of the characteristic becomes gradually flatter. The point on the characteristic where the current begins to level off is the knee-point of the characteristic, and is called the **pinch-off** voltage.

Referring to the set of characteristics depicted in Fig 8.3, we can say that the maximum current flows through the JFET when zero voltage is applied to the gate-to-source junction, and when the applied drain-to-source voltage is equal to, or greater than, the pinch-off voltage, V_p. However the voltage applied should be less than the breakdown voltage. This will be the operating region of the JFET for use as an amplifier. The maximum current, or the saturation current, when the gate-to-source voltage is zero, is called I_{DSS}, and is labelled as such in Fig 8.3. Thus, the static characteristic can be split-up in two regions with reference to drain-to-source voltage across the JFET, the operating region beyond the pinch-off voltage point (i.e. knee of the curve) and the ohmic region before the pinch-off voltage point. It should be recorded here that the JFETs biased in the ohmic region can be used as a voltage controlled variable resistance and the JFETs biased in the operating region, primarily, provide amplification functions.

It should also be noted that minimum drain current will occur when the gate-to-source voltage equals the pinch-off voltage, i.e., when $V_{GS} = V_p$. Theoretically, the minimum current is zero, but we can very well assume the minimum current to be about 1% to 2% of the current I_{DSS}.

The pinch-off voltage, V_p, for the JFET is the amount of gate-to-source voltage that will reduce the drain current to a negligible value as mentioned above. It should be clearly understood, however, that V_p is also the minimum drain-to-source voltage for the operation of the JFET as an amplifier. For instance, if V_p of the transistor is known to be about 4V, then the minimum V_{DS} value can not be lower than 4V, and, in fact, must be higher to reproduce a magnified image of the signal voltage. This is necessary since the pinch-off voltage line limits the movement of the operating point, producing distortion if the operating point tries to cross this point. The pinch-off voltage can be estimated as at the knee-point of the characteristic, or by finding the voltage V_{GS} at which the drain current reduces to almost zero; or it can be directly taken from the data-sheets provided by the manufacturers.

If we consider that the knee-point represents the pinch-off voltage, then joining the knee-points of each characteristic depicted in Fig 8.3 will produce a seemingly exponential curve, which we shall call a pinch-off voltage line, or simply, pinch-line.

The drain current, I_D, the saturation current, I_{DSS}, the gate-to-source voltage, V_{GS}, and the pinch-off voltage, V_p, are related by the following equation.

$$I_D = I_{DSS} \times (1 - V_{GS}/V_p)^2 \tag{8.1}$$

8.2 The Small Signal Parameters

Since $i_D = f(v_{DS}, v_{GS})$, hence, we can write for the JFET an equation representing the drain current as =

$$i_D = g_m \cdot v_{GS} + (1/r_d) \cdot v_{DS} \tag{8.2}$$

where the lower-case letters i and v represent the **change** in the respective current and the voltages.

From Equation 8.2, we can deduce the expressions for the values of

g_m = the transconductance, and r_d = the drain resistance.

For a constant voltage, V_{DS}, we have

$$g_m = \frac{i_D}{v_{GS}} \text{ mhos.} \tag{8.3}$$

and for a constant gate-to-source voltage, V_{GS},

$$r_d = \frac{v_{DS}}{i_{DS}} \text{ Ohms.} \tag{8.4}$$

The third parameter can be deduced from the first two. It is called **the amplification factor**, μ, and is defined as the ratio of the change in the drain voltage per unit change in the gate-to-source voltage for constant drain current.

Thus,

$$\mu = \frac{v_{DS}}{v_{GS}}$$

$$= \frac{v_{DS}}{i_{DS}} \times \frac{i_{DS}}{v_{DS}}$$

$$= r_d \times g_m \tag{8.5}$$

Equation 8.1 can be modified and approximated as

$$g_m = g_{mo} \times \left[1 - \frac{v_{GS}}{V_p}\right] \tag{8.6}$$

where, g_{mo} is the value of the g_m for $V_{GS} = 0$, and is given by,

$$g_{mo} = \frac{-2I_{DSS}}{V_p} \tag{8.7}$$

Equation 8.6 is an experimental one, and holds good for practical designs.

It will be useful to write the equation of g_m in a slightly different form.

From Equations 8.6 and 8.7, we have,

$$g_m = \frac{-2I_{DSS}}{V_p} \times \left[1 - \frac{V_{GS}}{V_p}\right]$$

Combining this with Equation 8.1, we get,

$$g_m = \frac{-2\,I_{DSS}}{V_p} \times \left[\frac{I_D}{I_{DSS}}\right]^{1/2}$$

$$= \frac{-2\,(I_D \times I_{DSS})^{1/2}}{V_p} \tag{8.8}$$

8.3 The Common Source Circuit

This circuit is similar to BJT common emitter circuit for its operation. This circuit provides a fairly high amplification. Figure 8.4 shows an *n*-channel JFET connected as a common source amplifier.

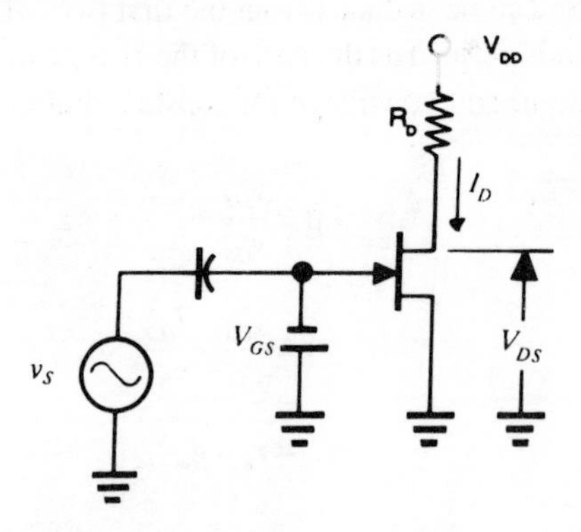

Fig. 8.4

The gate is biased negative by a separate gate voltage. As can be seen, the source terminal is common to, both, input and output. The resistance R_D, is the load resistance connected in the drain circuit. From the circuit, it is clear that the voltage drop across the resistance R_D, is equal to $I_D \times R_D$, and the voltage available across the FET is equal to V_{DS}, the value of which is given by the following equation written around the output circuit.

$$V_{DD} = I_D \times R_D + V_{DS} \tag{8.9}$$

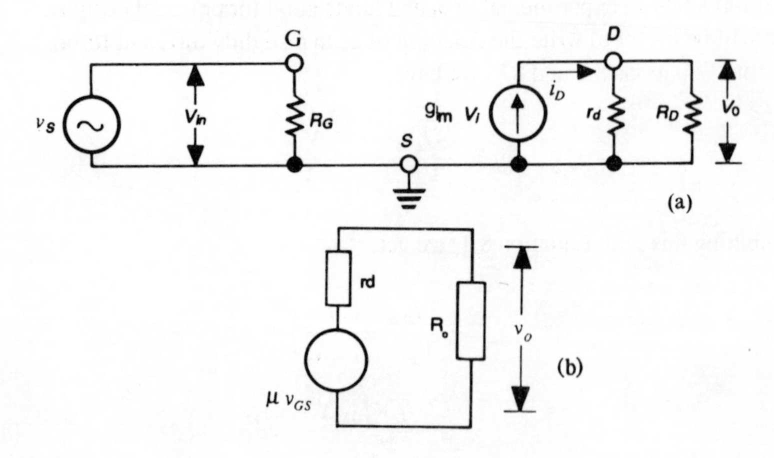

Fig. 8.5 (a, b)

Assume now that the signal voltage shown in the diagram has a peak magnitude of 0.1V, both in the positive and the negative cycles. This voltage will change the magnitude of the gate-to-source voltage by ±0.1 volts. The drain-to-source voltage will change according to Equation 8.9, since the change in the gate-to-source voltage is going to produce the change in the drain current, I_{DS}. Since, the signal voltage is assumed to be sinusoidal, the output voltage will also be sinusoidal, with the magnitude depending on the parameters and the load resistance connected in the circuit Figure 8.5 shows an equivalent circuit of the amplifier. Let us derive some equations which will help us select various components when we attempt the design of the circuit.

8.4 Voltage Gain

Since we are dealing with incremental values, we shall use lower case letters for the same, as in the case of BJT amplifiers.

$$\text{The output voltage } = i_d \times \frac{r_d \times R_D}{r_d + R_D} \tag{8.10}$$

Also since $\quad i_d = -g_m \times v_i$

$$v_o = -g_m \times v_i \times \frac{r_d \times R_D}{r_d \times R_D}$$

from which, the voltage gain $= A_v$

$$= \frac{v_o}{v_i} = -g_m \times \frac{r_d \times R_D}{r_d + R_D} \tag{8.11}$$

Since in many cases we shall be taking the value of load resistance to be very small as compared to the drain resistance, r_d, the above equation may be rewritten as

$$A_v = \frac{-g_m \times r_d \times R_D}{r_d}$$

$$= -g_m \times R_D \tag{8.12}$$

8.5 Self-biased JFET Amplifier

Since the gate should be kept negative with respect to source-terminal of the FET (assuming *n*-channel JFET), one of the methods to achieve the same is called self-biased circuit shown in Fig. 8.6. The current I_D flowing through R_S creates a voltage drop equal to $I_D \times R_S$. The source is pulled-up by this voltage. Since the gate is grounded through resistance, R_G, the voltage at the gate, V_G, is equal to zero. The gate voltage with reference to the source, therefore, is negative by a magnitude equal to $(I_D \times R_S)$.

The equivalent circuit for the common-source amplifier is shown in Figs 8.5(a) and (b).

Figure 8.5(a) shows the current equivalent circuit, whereas Fig 8.5(b) shows a voltage equivalent circuit. (The resistance R_S is not shown in the diagram.)

From Fig 8.5(a), we have

$$v_o = i_2 \times \frac{r_d \times R_D}{r_d + R_D}$$

But since $\quad i_2 = g_m \times v_g$

we get $\quad v_o = g_m \times v_g \times \frac{r_d \times R_D}{r_d + R_D}$

Rearranging, we get

$$A_v = \frac{v_o}{v_g} = g_m \cdot \frac{r_d \times R_D}{r_d + R_D} \tag{8.13}$$

Also from Fig 8.5(b), we get

$$v_o = \mu\, v_g \frac{r_d}{r_d + R_D}$$

Hence, $\quad A_v = \frac{v_o}{v_g} = \frac{\mu R_D}{r_d + R_D} \quad (8.14)$

Equations 8.13 and 8.14 have been arrived at after neglecting the effect of resistance, R_S, which reduces the output voltage, and hence the gain. The effect, normally, is negligibly small, since resistance R_S is effectively reduced to a value equal to $(g_m \times R_S)$. The resistance R_S is usually shorted by a capacitor, C_S, for amplifier applications. However, where not 'shorted', the gain Equation 8.13 has to be modified to

$$A_v = g_m \times \frac{R'_D}{1 + g_m \times R_S} \tag{8.15}$$

where R'_D is the parallel combination equivalent of R_D and R_L, if considered present.

Let us analyse such a circuit by taking a numerical problem.

EXAMPLE 8.1

Refer to Fig 8.6. The numerical values are shown therein.

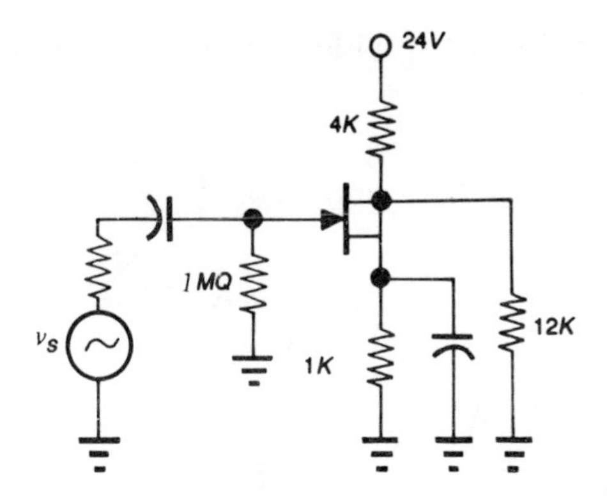

Fig. 8.6

The following data is given:

$$V_{DD} = 24 \text{ V}$$
$$V_p = 6\text{V}$$
$$V_{GS} = -3\text{V}$$
$$I_{DSS} = 12 \text{ mA}.$$

Solution

From Equation 8.1, we have

$$I_{DS} = I_{DSS}\left[1 - \frac{V_{GS}}{V_P}\right]^{1/2}$$
$$= 12\text{mA } (1 - 3/6)^{1/2}$$
$$= 3\text{mA}.$$

The voltage across the resistance, R_s, $= 3\text{mA} \times 1k$
$= 3\text{V}$

This is also the voltage available at the source terminal.

Since the gate is grounded through the resistance, R_G, the voltage at the gate is equal to zero. Hence, the gate-to-source voltage, $V_{GS} = -3\text{V}$.

Likewise, the voltage across the resistance, R_D, is given by ($I_D \times R_D$)

$$= 3\text{mA} \times 4k$$
$$= 12\text{V}$$

The drain-to-source voltage, V_{DS}, is, therefore, given by

$$= V_{DD} - (I_D \times R_D) - V_{RS}$$
$$= 24 - (3\text{mA} \times 4k) - 3$$
$$= 9\text{V}$$

The figure shows the *dc* voltage distribution for the circuit. From Equation 8.8, we have

$$g_m = \frac{2}{V_p}(I_D \cdot I_{DSS})^{1/2}$$

$$= (2/6) \times (3\text{mA} \times 12\text{mA})$$
$$= 2 \text{ mmho.}$$

From Equation 8.15, we have

$$A_v = g_m \times \frac{R'_D}{1 + g_m \times R_S}$$

where R'_D is a parallel combination of two resistances, R_D and R_L.

$$= 2\text{mmho} \times (4\text{k} \,//\, 12\text{k}) \,/\, (1 + 2\text{mmho} \times 1\text{k})$$
$$= 2$$

Hence if the input *ac* voltage happens to be 2V p-p, then the output voltage will be 4V p-p.

8.6 Biasing for Zero Current Drift

Figure 8.7 indicates the effect of the temperature on transfer characteristics. This shows that if the temperature increases, then for the same value of the gate-to-source voltage, V_{GS}, the value of the I_{DS} **decreases**. As can be seen, the cut-off voltage also increases. If the temperature decreases, the reverse phenomena takes place. I_{DS} increases for the same value as V_{GS}. These characteristics are shown in the figure. This would mean, then, that the biasing implemented may not work if the working temperature increases. There is, however, one redeeming feature in Fig 8.7, and that is that there is one point available in this set of characteristics where the temperature does not seem to have any effect. This point is the cross-over of all characteristics, as shown in the diagram.

If we, therefore, bias the JFET precisely at this point, then there will be no noticeable drift in the Q-point. At any other point on the characteristic, the drift of I_D is of the order of 0.7% / °C.

However, like in a bipolar transistor, the increase in the temperature should provide higher current, I_D, since the gate-to-source source voltage is going to be affected. It has been found that the increase in I_D is equivalent to a change of 2.2mV /°C in the magnitude of V_{GS}.

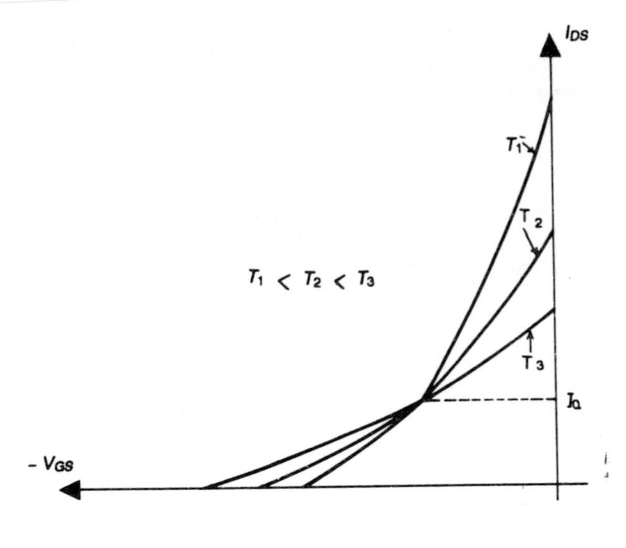

Fig. 8.7

Keeping the above two effects in mind, we can write the condition for zero drift as

$$0.007\, I_D = 0.0022 \times g_m \tag{8.16}$$

giving

$$V_P - V_{GS} = 0.63 \text{ V} \tag{8.17}$$

from equations 8.1, 8.6, and 8.7. (Only the magnitudes of the voltages should be considered and not the signs.)

Equations 8.1 and 8.6 will, therefore, be modified for zero drift as

$$I_D = I_{DSS}\left(\frac{0.63}{V_P}\right)^2 \tag{8.18}$$

and,

$$g_m = g_{mo}\,\frac{0.63}{/V_P/} \tag{8.19}$$

8.7 Biasing against device variation

Normally information is provided by the manufacturers for the amount of possible variation to be expected in the devices in the current, I_{dss}, and the pinch-off voltage, V_p. The maximum and minimum values that can be expected for room-temperature are given in the data sheet. These set of values, when plotted, produce the set of characteristics as shown in Fig. 8.8.

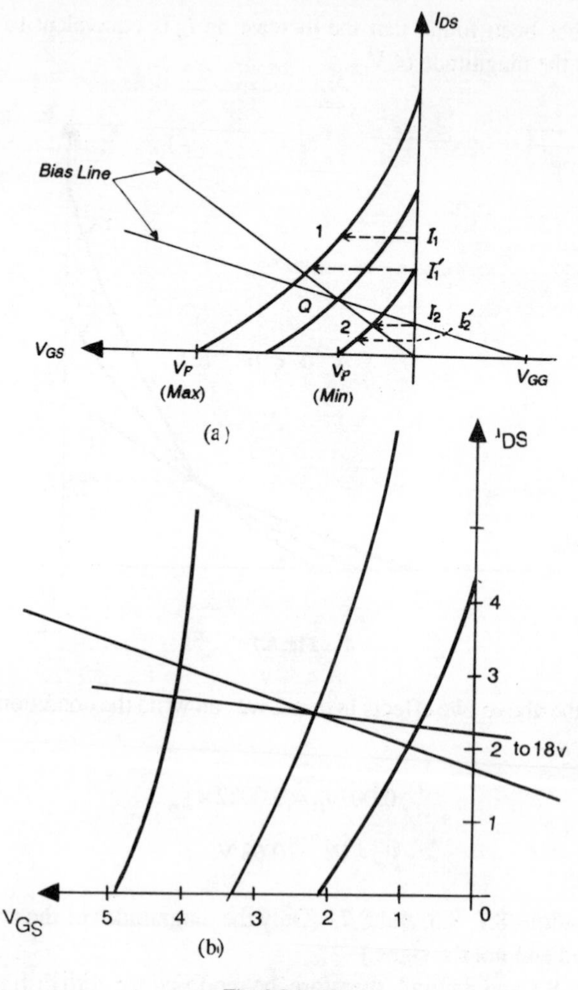

Fig. 8.8 (a, b)

As shown there in, the top characteristic refers to the maximum values and the bottom curve is for the minimum values. Suppose it is desired that the quiscent point should not drift beyond I_1, and remain below the point I_2, then the bias line $V_{GS} = I_{DS} \times R_S$, should intersect the characteristics between points 1 and 2 as shown in the figure. If it does, then the point of operation will not venture beyond these points and the change in the device will not affect the operation unduly.

Consider the second bias line drawn in the figure. It can be seen that the conditions imposed are such that the bias line drawn between points for maximum and minimum values of the currents I'_1 and I'_2 does not pass through the origin; instead it cuts the x-axis at another point V_{GG}. In the previous case the bias-line satisfied the equation $V_{GS} = I_{DS} \times R_S$, which will now change to

$$V_{GS} = V_{GG} - I_{DS} \times R_S$$

Such a bias line may be obtained by adding a fixed bias as shown in Fig. 8.9a, or by a method shown in Fig 8.9b. It is interesting to note the third possibility, viz. the imposed conditions make the bias-line to intersect the *x*-axis in the negative part. (Not shown in the diagram.) In such a case, however, the bias will have to be provided by a separate supply.

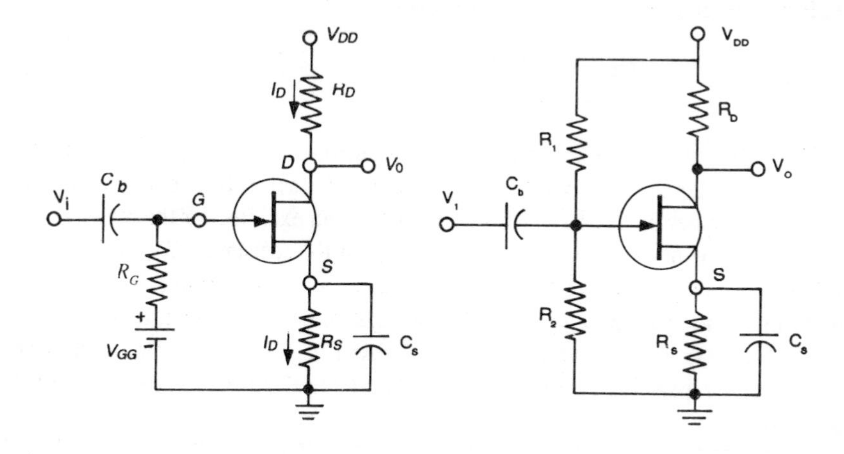

Fig. 8.9 (a and b)

EXAMPLE 8.2

Design a "CS" amplifier, single stage using JFET having the characteristics as shown, with biasing circuit to provide stability of operating Q-point against device variation so that I_{DS} varies between 2 mA and 3.0 mA to provide $A_v = -15$. Assume $r_d = 20$ K-ohm for the FET.

Solution

Refer the characteristics drawn in the figure 8.8(a). Plot the limits of I_{SD} as shown in the graph at points A and B and draw the bias line [i.e. join $I_{D(\max)}$ and $I_{D(\min)}$]. Extend the line so that it cuts the X-axis at VGG. Or alternatively, draw the line in between these two points as shown in the figure 8.8(b). Here the line is drawn with the following co-ordinates

(0V, 2.25 mA) and (−4V, 2.75 mA).

The slope of this line determines the resistance R_S.

$$\text{Slope of the bias line} = \frac{Y}{X} = \frac{2.75 - 2.25}{4.0 - 0}$$

$$= 0.125 \times 10^{-3}$$

$$\text{Now slope} = \frac{1}{R_S}$$

Therefore, $R_S = 8$ K ohms.
We select $R_S = 8.2$ K, the nominal value available.

From the graph (typical), the Q point is at

$$I_{DQ} = 3.6 \text{ mA.}$$
$$V_{GSQ} = -0.9 \text{ V}$$

Once the value of the resistance, R_S, is determined and the Q-point is marked on the characteristic (the typical values characteristics), it is a routine procedure to calculate other parameters. This part is left as an exercise for the readers.

It should be noted, however, that the value of the resistance, R_S, calculated as above could be different if the bias line slope selected is different. For instance, if the bias line were drawn joining the two extreme values given, then the slope would be $(1/3) \times 10^{-3}$ or the resistance, R_S, will be 3 k ohms. The readers are suggested to repeat the rest of the calculations and find the difference in the results obtained.

EXAMPLE 8.3

Design a single stage amplifier employing FET type 2N3822 in CS configuration to give Voltage gain $A_v = -9$, and *ac* output of 2.5 Voltage rms with biasing circuit to give:

(i) Zero temperature drift

and (ii) to operate FET at $I_d = \frac{I_{dss}}{2}$

Specifications of 2N3822:

$$V_{DS(\max)} = 50 \text{ Volts}$$
$$V_{DG(\max)} = 50 \text{ Volts}$$
$$V_{GS(\max)} = 50 \text{ Volts}$$
$$P_{O(\max)} = 300 \text{ mW (at 25°C)}$$
$$I_{DSS} = 2\text{mA}$$
$$G_{mo} = 3000 \text{ u-mhos.}$$
$$V_p = 6\text{V}$$
$$r_d = 50 \text{ K}$$

Solution for (i)

In FET, the drain current decreases with the increase in temperature due to the decreased mobility of the majority carrier, which is due to the increased vibration of ions and atoms. The reduction in I_D is 0.7%/°C.

Further, the width of gate-to-channel barrier decreases with increase in temperature. This allows I_D to increase, and increase in I_D is found to be equivalent to a change of 2.2 mV/°C in V_{GS}.

Therefore, at balance, the condition for zero temperature drift is

$$0.007\, I_D = 0.0022\, g_m \tag{1}$$

Also, we know that

$$I_D = I_{DSS}\left(1 - \frac{V_{GS}}{V_p}\right)^2 \tag{2}$$

$$g_m = g_{mo}\left(1 - \frac{V_{DS}}{V_p}\right) \tag{3}$$

$$g_{mo} = \frac{-2 \times I_{DSS}}{V_p} \tag{4}$$

From equations 3 and 4, we have

$$g_m = \frac{-2 \times I_{DSS}}{V_p} \times \left(1 - \frac{V_{GS}}{V_p}\right) \tag{5}$$

From equations 1, 2 and 5, we get

$$0.007\left(1 - \frac{V_{GS}}{V_p}\right)^2 = 0.0022\left(-2\frac{I_{DSS}}{V_p}\right) \times \left(1 - \frac{V_{GS}}{V_p}\right)$$

From which, $\quad V_p - V_{GS} = -0.63\text{ V}$

$$/V_p/ - /V_{GS}/ = 0.63\text{ V}$$

For Zero temperature drift,

$$V_{GS} = V_p - 0.63$$

$$V_{GS} = 6 - 0.63$$

$$V_{GS} = 5.37\text{V}$$

We know $$I_D = I_{DSS} \times \left(1 - \frac{V_{GS}}{V_p}\right)^2$$

Therefore $$I_D = (2 \times 10^{-3})\ \{1 - (-5.37/-6)\}^2$$

or $$I_D = 2.205 \times 10^{-5}\ \text{Amps}$$

Selection of R_G

R_G depends on Z_{in} required.

$$Z_{in} = R_G$$

Select $$R_G = 1\ \text{M}$$

Selection of R_S

As I_G is negligibly small; $V_{RG} = 0$
[Since no current passes through the resistance R_G, $V_{RG} = 0$]
Therefore, $V_G = 0$ (i.e. the gate terminal is grounded)

Hence, $$V_{RS} = -V_{GS} = -5.37\ \text{V}.$$

From which $$R_S = \frac{V_{RS}}{I_D}$$

$$= 5.37\text{V} / 22.05\ \text{uA}$$
$$= 243.53\ \text{K}$$

We select $R_S = 220\text{K}$, the nearest available standard value.

Selection of R_D

Voltage Gain $$A_v = -g_m \times R_{Deff}$$

where $$R_{Deff} = \frac{R_D \times r_d}{R_D + r_d}$$

Also $$g_m = g_{mo}\left(1 - \frac{V_{GS}}{V_p}\right)$$

Therefore, $$g_m = 3000 \times 10^{-6}\ \{1 - (-5.37/-6)\}$$

Hence, $$g_m = 3.15 \times 10^{-4}.$$

$$R_{Deff} = \frac{A_v}{-g_m} = \frac{-9}{-315 \times 10^{-6}}$$

$$R_{Deff} = 28.571\text{ K}$$

But, $$R_{Deff} = \frac{R_D \times r_d}{R_D + r_d}$$

Therefore, $$28.571 = \frac{R_D \times 50}{R_D + 50}$$

From which, R_D = 66.66K

We select R_D = 68K the nominal value.

Power dissipation in $R_D = (I_D)^2 \times R_D$

$= (0.02205 \times 10^{-3})^2 \times (60 \times 10^3)$

$= 0.0033$mw.

We select R_D = 68 K, 1/8W.

NOTE:

With R_D = 68K,

$V_{RD} = I_D \times R_D$

$= (0.02205 \times 10^{-3})\ (68 \times 10^3)$

= 1.5 Volts.

This is less than the peak voltage = 3.53V required. Hence, we will have to increase the resistance R_D.

Selection of V_{DD}

$V_{o(peak)} = (V_{o(rms)} \times /2.$

$= 2.5 \times /2$

= 3.53 Volts.

There are three resistances in the circuit, therefore, in the worst case, we should add tolerences of all the resistances and consider them to affect the circuit in the worst possible way.

Hence, $V_{RD} = 1.3 \times V_o\text{(peak)}$

= 4.589 Volts

Now with this value of V_{RD}, we calculate the resistance R_D again.

$$R_D = \frac{4.6}{0.02205} = 208.61\text{K}$$

$$R_D = 220\text{K. (say)}$$

Power dissipation $= (0.02205)^2 \times (220 \times 10^3)$

$= 0.106$ mW.

We select $R_D = 220\text{K}, 250\text{mW}.$

$$V_{DSQ} = V_p + V_o$$

$$= 6 + 4.6$$

$$= 10.63 \text{ Volts.}$$

$$V_{DD} = V_{RD} + V_{DSQ} + V_{RS}$$

$$= 4.6 + 10.6 + 5.37$$

$$= 20.57 \text{ Volts}$$

We select $V_{DD} = 21$ Volts

Selection of C_s

At the lowest operating frequency 15Hz (say)

$$X_{cs} = \frac{R_S}{10}$$

$$= \frac{220 \times 10^3}{10} = 22\text{K}$$

Therefore, $$C_s = \frac{1}{2\pi \times 15 \times 22\text{K}}$$

Giving, $C_s = 0.48$ mf. (Voltage rating of 10V)

Solution for (ii)

$$I_{DS} = \frac{I_{DSS}}{2}$$

$$= 2\text{mA}/2 = 1\text{mA}.$$

Now,

$$I_{DS} = I_{DSS}\left(1 - \frac{V_{GS}}{V_p}\right)$$

From which $V_{GS} = 1.8\text{V}$

Also $$g_m = g_{mo}\,(1 - \frac{V_{GS}}{V_p})$$

Giving, $$g_m = 3000 \times 10^{-6}\,\{1 - (-1.8/-6)\}$$
$$= 2100\ \mu\text{-mhos}$$

Selection of R_S

As I_G is considered as zero, $V_{RG} = 0$ and therefore, $V_G = 0$.

Therefore, $$V_{RS} = -V_{GS}$$
$$= -(-1.8\text{ V})$$
$$= 1.8\text{ V}$$

$$R_S = \frac{1.8}{1 \times 10^{-3}} = 1.8\text{ K.}$$

Power dissipated in $$R_S = (1 \times 10^{-3})^2\,(1.8 \times 10^3)$$
$$= 1.8 \times 10^{-3}\text{ W.}$$

We select $$R_S = 1.8\text{K, 1/8W.}$$

Selection of R_G

R_G depends on Z_{in} required.

$$Z_{in} = R_G$$

We select $$R_G = 1\text{M Ohm}$$

Selection of C_s

At the lowest frequency, 15 Hz (say),

$$X_{cs} \ll R_S$$

We assume $$X_{cs} = (1/10) \times R_S = 180\text{ Ohm.}$$

$$C_s = \frac{1}{2\,\pi \times 15 \times 180},$$

giving, $$C_s = 58.94\ \mu\text{f.}$$
$$C_s = 68\ \mu\text{F (Std value).}$$

Selection of R_D

$$A_v = -g_m\,(r_D \,//\, R_D)$$

$$A_v = -g_m \times R_{Deff}, \text{ where, } R_{Deff} = \frac{R_D \times r_d}{R_D + r_d}$$

$$-9 = -2100 \times 10^6 \times R_{Deff}$$

From which, $R_{Deff} = 4.28\text{K}$

Hence, $4.28\text{K} = \dfrac{R_D \times 50\text{K}}{R_D + 50\text{K}}$

From which, $R_D = 4.68 \text{ K}$

We select $R_D = 4.7\text{K}.$

Selection of V_{DD}

$$v_o = 2.5 \text{ V}$$

$$v_{o(peak)} = 2.5 \times \sqrt{2}$$

$$= 3.53 \text{ volts}$$

$$v_{op} = 3.53 \times 1.3 = 4.6 \text{ V},$$

taking into consideration the tolerences (of three resistances used).

To find R_D again

$$R_D = \frac{4.6}{1 \times 10^{-3}}$$

$$R_D = 4.6 \text{ K}$$

We select $R_D = 4.7\text{K}, 1/4\text{W}.$

$$V_{DSQ} = V_{o(p)} + V_p$$

$$= 4.63 + 6$$

$$V_{DD} = V_{RD} + V_{DSQ} + V_{RS}$$

$$= 4.7 + 10.63 + 1.8 = 17.1\text{V}.$$

We select $V_{DD} = 18\text{V}.$

EXAMPLE 8.4

If two such identical stages are to be coupled by R-C coupling, calculate the value of the coupling capacitor to give 3 db frequency response upto 15Hz [Consider any one of the above two cases in part (i) or part (ii)].

We consider that the two stages are of part (i) type of the above example.

The output impedance $= r_D \,//\, R_D$

$= 50\text{K} \,//\, 220\text{K}.$

$= 44\text{K} = R_{o1}$

The input impedance of the second stage =

$$R_{in2} = R_{G2} = 1\text{M Ohm}$$

the value of the coupling capacitor required to give 3 db frequency response at 15 Hz is given by

$$C_C = \frac{1}{2\pi f\,(R_{o1} + R_{in2})}$$

$$C_C = \frac{1}{2\pi \times 15 \times (44\text{K} + 1000\text{ K})}$$

$$= 10.1 \text{ nf.}$$

We select $C_C = 22\ nf$, which is the nearest higher standard value available.

9

POWER AMPLIFIERS

In applications like speech or music reproduction, or in public address systems, a large amount of power is required to be output to the load. This means that the transistors must be operated over almost the entire length of the transfer characteristic. Apart from the large excursion on the characteristic, it is also of paramount importance to have low distortion in the output. This normally means that the amplification process first involves one amplifier where, essentially, voltage amplification is done, and then the amplified signal is fed to the above mentioned amplifier where the emphasis is more on power amplification than the voltage.

Since this type of amplifier requires using the movement of the operating point over the entire length of the load line, care should be taken to avoid distortion because of the non-linearity of the characteristics. The second point that should be kept in mind is the power dissipation of the device. The power efficiency of the device should be kept high. In trying to achieve high efficiency, we will have to resort to several empirical formulae or assumptions while selecting supply voltages, or choosing a load resistance, for that matter.

9.1 Power Amplifiers

Like voltage amplifiers discussed earlier, power amplifiers also can be classified to have three operating classes, viz., class A, class B and Class C.

In class A operation, the collector current flows all the time, resulting in low power efficiency. This mode is selected, however, when low distortion is desired.

Class B operation provides greater power efficiency and higher power output, but also results in higher distortion. Since only half cycles are amplified, and it generates very high amount of harmonics. This can, however, be taken care of by using push-pull operation.

Class C operation leads to extremely high distortion. Though the efficiency is relatively very high, this kind of amplifiers are generally not preferred for linear operation.

9.2 Transformer Coupled Power Amplifier

When higher output power is the consideration, normally, load resistance as employed in the voltage amplifier, is not used. This is because the *dc* current component flowing in the circuit causes an unacceptable large amount of power loss. Also since the power output required is high, for a given voltage variation in the output, the value of the load resistance required is fairly low. Therefore, in most cases, a transformer connected load is used. Refer to Fig. 9.1.

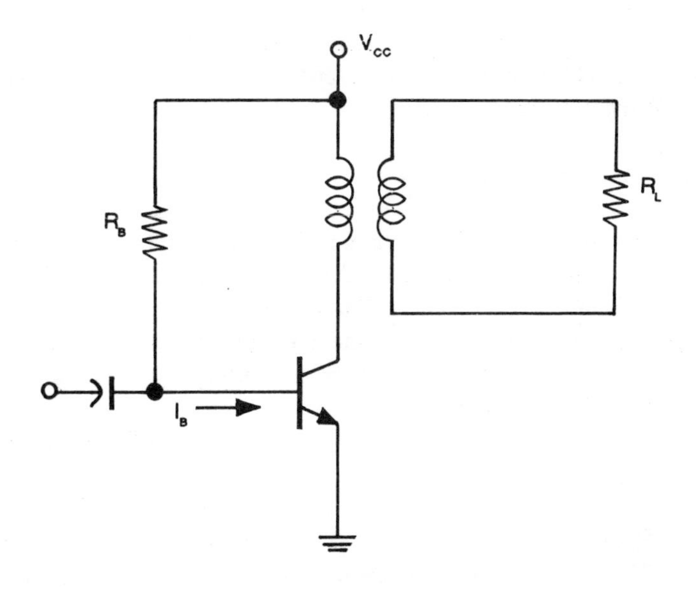

Fig. 9.1

The transformer primary offers a low *dc* resistance; at the same time *ac* resistance can be adjusted by proper selection of the resistance R_L, or by selecting a proper turns ratio for the transformer.

For a transformer the voltage and current are related by a turns ratio, ***k***, which is given by

$$k = \frac{V_p}{V_s} = \frac{I_s}{I_p} = \frac{n_1}{n_2} \tag{9.1}$$

The impedance as seen by the primary of the transformer is given by

$$Z_p = \frac{V_p}{I_p} = \frac{k\,V_s}{I_s/k} = \frac{k^2\,V_s}{I_s} = k^2 Z_s \tag{9.2}$$

where Z_p is equivalent impedance of the transformer when referred to primary. This means that a load resistance of a value R, if connected to the secondary, gets multiplied by the square of the turns ratio, and this value appears as an *ac* resistance in the primary circuit which, in our case, happens to be the collector resistance.

For example, if the turns ratio is 10, and the resistance is of 10 ohms, then the equivalent value appearing in the primary circuit is 1000 ohms. We are, however, assuming that the transformer is ideal. The ideal transformer has no leakage inductance and no winding resistance; thus it produces only voltage transformations and does not introduce any phase-shift between the primary and the secondary signals. Such a transformer, however, is not possible. The resistance of the winding also is quite comparable, and we shall, therefore, take them into account while designing the amplifiers.

9.3 Class-A Power Amplifiers

Keeping in mind the allowable distortion and the maximum power dissipation of the device, one would like to design for the maximum power output. It should be kept in mind that the selection of the Q point also decides the output obtainable, as also the supply voltage selection. The maximum supply voltage that can be selected will largely depend upon, and be limited by, the maximum collector voltage rating of the transistor.

The Q point selection will also decide the distortion available in the output signal.

Also we have to keep in mind that the *dc* power input to the device totals the *ac* power output, device dissipation power and the power loss in the transformer resistance. The power dissipation of the transistor is largely due to the product of the collector-emitter voltage and the collector current, i.e. $V_{CE} \times I_C$.

Assuming that losses which occur in the transformer winding resistance and in the biasing network are small, we can say that the device dissipation equals total input power less transformer *ac* power output. In the absence of a signal, i.e., at a Q-point operation, the device dissipation is maximum in class-A operation. While designing such amplifiers, therefore, this should be kept in mind. First, normally, a dissipation curve should be drawn on the output characteristics as shown in Fig. 9.2. This can be drawn from the relation:

$$\text{The device dissipation } = V_{CE} \times I_C \qquad (9.3)$$

Since the maximum dissipation occurs at quiscent point, both V_{CEQ} and I_{CQ} should be taken into account. Any operating point on the load line which falls below this locus is a safe operating point. For large outputs, the load line should be drawn tangential to this power curve, and the inverse of the load line thus drawn should be taken as the load resistance.

Once the Q-point has been decided, as shown in Fig. 9.2, the maximum swing of I_C can be determined. Traversing up the line, till the load line comes near to the 'knee' portion of the characteristics, we find a point beyond which the operating point can not move. This excursion, at least theoretically, is the possible swing in one direction, or is equal to $V_s/2$, where V_s is the voltage swing possible.

Let us consider one example. Refer to the output characteristic shown in Fig. 9.2. The transistor for which the characteristics are given, has maximum power dissipation $P_{d(\max)} = 13.5$ watts and $V_{CE(\max)} = 30$ V.

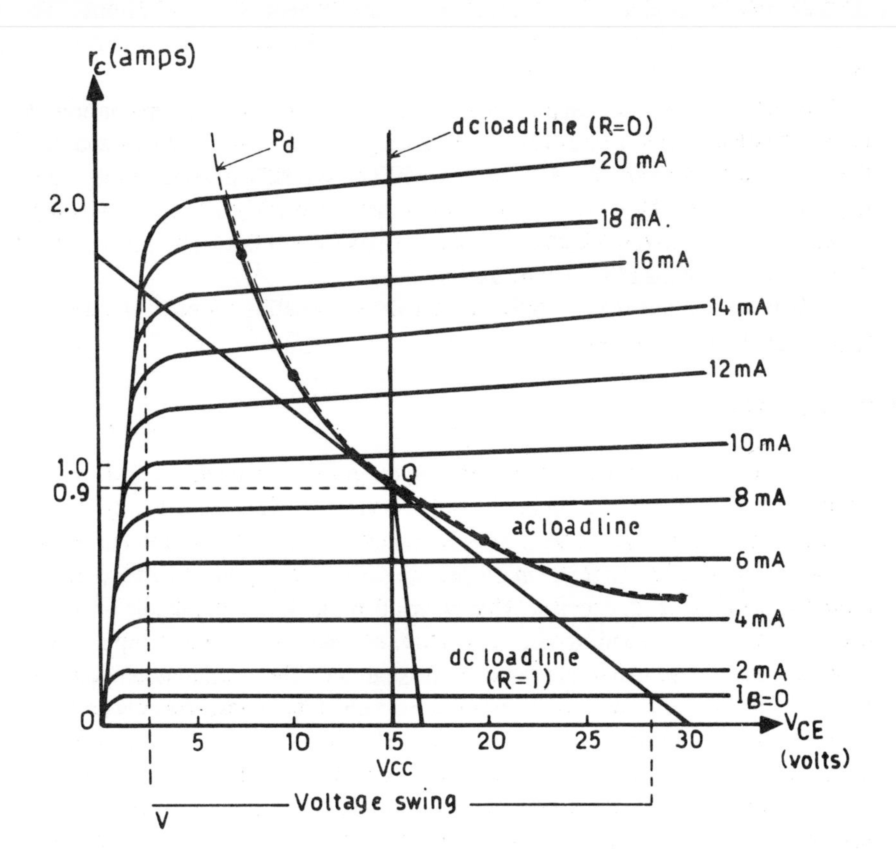

Fig. 9.2

Let us draw an *ac* load line. Since maximum V_{CE} is 30 volts, we shall draw a tangent from this point to the power dissipation curve. This line when extended cuts the Y-axis at 1.8 A point. The point at which the load-line is tangent to the power dissipation curve is a Q-point, which from the figure is (15 V, 0.9 A). The Q-point dissipation is obviously maximum, and is 13.5W at 25°C. Now we have to take some arbitrary decisions for determining the maximum swing of the signal.

As shown earlier, the swing can be taken to the extent where the load line meets the curved portion of the characteristics. At this point, V_{CE} is 2.5 volts. Hence, we can assume that the output voltage can swing from 15 Volts down to 2.5 V on one side. On the other side, the swing seems to be a little larger, limited only by the zero-base-current characteristics. We shall assume symmetry on the other side. Thus, the swing on the other side will be from 15 Volts to (15 + 12.5)V = 27.5 Volts. This means that the input voltage signal will force a change of base-current from 18 mA to about 2 mA, with a Q-point base-current of about 9.0 mA. To make it symmetrical, we may permit the base-current to swing from 2.0 mA to 16 mA, i.e., 7 mA in either direction.

This corresponds to the output voltage swing from 27 V to 3V. This limit will reduce the distortion to a great extent. Assuming that the signal is absent and that the transformer is ideal, the *dc* load line will be vertically passing through the Q-point as shown. However, if we assume that the transformer has a winding resistance (*dc*) of, say, 0.5 Ohm, and that resistance R_E of 0.5 Ohm is also introduced, then the *dc* load line will be as shown.

It should be noted that, in practice, it will be more appropriate to take the Q-point at about 60% to 80% of V_{cc}.

9.4 Derating

At this point we should also be considering the power dissipation and the temperature effect. The maximum permissible power dissipation is normally specified by the device manufacturer at a certain temperature, either the ambient or the case temperature. For example, in the previous case it was given as 13.5 watts at room temperature of 25 degrees. This means that the device is capable of dissipating 13.5 watts at room temperature only, and not at a higher temperatures which evidently will result due to power dissipation. The maximum dissipation capability of the device will reduce as the temperature increases. This decrease in the maximum power dissipation capability at elevated temperature is called **derating.** The manufacturers specify the **derating factor** in terms of Watts / °C. If in the above case, if the derating factor had been 0.1 W/ °C, and the temperature had been 50°C, then the maximum power that a device can dissipate will reduce to

$$13.5\text{W} - (50 - 25) \times 0.1\ \text{W}$$
$$= 13.5\ \text{W} - 2.5\ \text{W}$$
$$= 11\ \text{W}.$$

The load lines will have to be accordingly shifted to account for this.

In practice, however, it is safe to assume that the transistor is capable of delivering only half, or even less, power than what it is rated for. Heat-sinks will be of great help and almost invariably will have to be used for large power outputs.

9.5 Power Relations in Transformer-Coupled Class-A Amplifiers

As already seen, the load resistance, R_L, connected to the secondary of the transformer is reflected in the primary of the transformer as R_p, and is given by

$$R_p = k^2 \times R_L$$

where k is the turns ratio of the transformer.

The resistance, R_p, is an equivalent *ac* resistance, and hence, is numerically equal to the inverse of the slope of the *ac* load line. As shown in Fig. 9.3, the *dc* load line is vertical, as we have assumed a negligible primary winding resistance. The Q-point is at V_{cc}. To force the operating point at quiscent value, we shall have to use a conventional method of biasing such that the Q-point collector current, I_{CQ}, flows through the transistor. With this arrangement, the maximum swing of the operating point in either direction is going to be equal to V_p, and we shall obtain a peak-to-peak variation of the voltage equal to $2V_p$. Since the voltage V_p is equal to the supply voltage V_{CC}, it means that the total variation available equals twice the supply voltage. This is the point where the *ac* load line intersects the X-axis. The intersect on the Y-axis will give us the maximum current flowing, namely, $I_{C(max)}$.

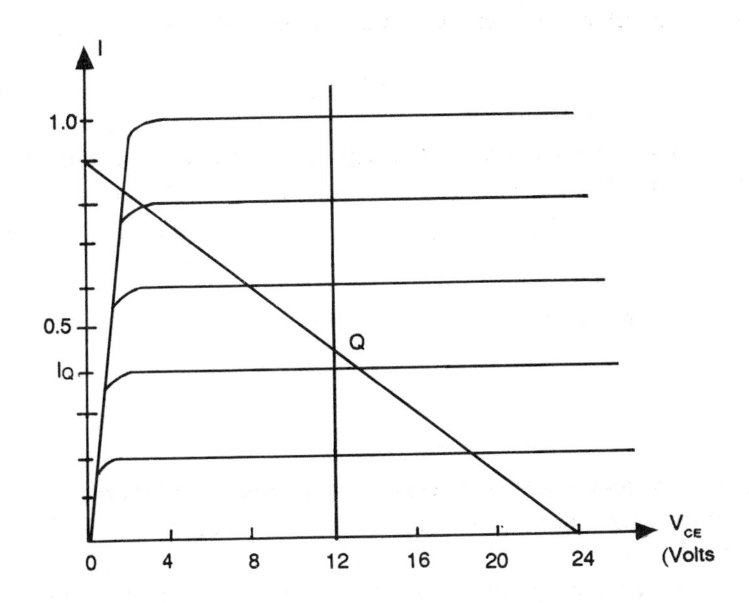

Fig. 9.3

It should be noted that since the primary winding resistance is assumed to be zero, there is no limit of the maximum current that will flow through the transistor. In practice, however, the maximum current should not exceed the permissible value of the current, nor should it be large enough to cause magnetic saturation of the transformer core. Like the voltage, the current at the quiscent point is half of the maximum, giving total peak-to-peak current swing of $2I_Q$.

One important point that should be kept in mind is that, unlike resistance load in the collector circuit, this kind of transformer connection provides a swing that **exceeds the supply voltage.** Hence, the transistor should have a collector breakdown voltage which is at least twice the supply voltage.

The *ac* power delivered to the load is

$$P_o = \frac{\left(V_{2p}/\sqrt{2}\right)^2}{R_L} \tag{9.4}$$

where V_{2p} is the peak value of the output, or the load voltage swing.

The *dc* power input to the circuit

$$P_{in} = V_{cc} \times I_{CQ} \text{ watts.} \tag{9.5}$$

We can, therefore, calculate the power efficiency of the amplifier as

$$= \eta = P_o / P_{in} \tag{9.6}$$

Since the primary maximum swing is going to be equal to the supply voltage, V_{cc}, we can deduce the secondary winding voltage or the load voltage swing as

$$= V_{cc} \times k \tag{9.7}$$

Again the load resistance is the inverse of the slope of the load line, we have,

$$I_{CQ} = \frac{V_{CC}}{R_L} \tag{9.8}$$

Thus, the maximum possible efficiency for transformer-coupled class A amplifier as found from Equations 9.6, 9.7 and 9.8 is equal to 0.5, or 50%.

In practice, however, full voltage swing of twice the supply-voltage can not be achieved, nor can it supply twice the I_{CQ}, since at the higher end, the transistor goes into saturation.

EXAMPLE 9.1

Design a class A power amplifier to output 2W to the speaker of 4 ohms. Assume the supply voltage available to be = 12V.

Solution

Refer to the Figs 9.3 and 9.4

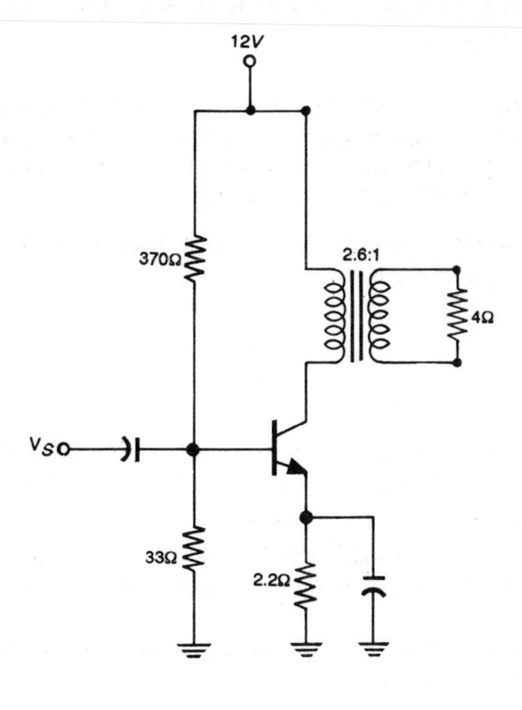

Fig. 9.4

1. Choice of Transistor

Since the maximum efficiency available in class A amplifier is 50%, we shall choose a transistor with a power rating of at least 3 to 4 times the output power required.

From the list of the available transistors, we can select ECN149 transistor, which has maximum power dissipation, $P_{d(max)} = 30W$, at 25°C and $I_{C(max)} = 4A$. The saturation voltage, $V_{CE(sat)} = 1.0$ V.

(It should be noted here that any other transistor available with the maximum power dissipation capability of about 10 Watts would be suitable)

2. Load Resistance

$$R_L = \frac{(V_{CC} - V_{sat})^2}{2 \times P_{o(max)}}$$

$$= \frac{(12 - 1.2)^2}{2 \times P_{o(max)}}$$

$$= 27 \text{ Ohms.}$$

It should be noted that this resistance is the equivalent resistance in the collector circuit of the transistor.

[**Note:** $P_{o(max)}$ in the above formula should be taken at about 10% more than the required value, as some power is going to be wasted in resistance of the transformer, etc. Also, since V_{sat} = 1V, the actual value of the saturation voltage taken should be slightly greater to reduce the chances of possible distortion at the knee point.]

Since the equivalent resistance of the transformer secondary referred to the primary is $= r_p = r_s \times K^2$, we have,

$$K = (27 / 4)^{1/2}$$

$$= 2.6$$

From the *ac* load line shown in Fig. 9.3, we can calculate the quiscent collector current $= I_c = 12 / 27 = 0.44$ A, at a point $V_{CC} = 12$ V, $R_L = 27$ Ohms.

The collector voltage can swing, on the lower side, upto $V_{sat} = 1$V, (say, 1.5 V). On the other side, it can swing, therefore, upto 22.5 V. The current can have a swing of, typically, upto about 0.8A (instead of 0.88 A, which is one of the load-line co-ordinates).

This swing in the voltage and the current, therefore, will provide us with the actual power delivered as

$$P_o = \frac{V_{pp}}{2 \times \sqrt{2}} \times \frac{i_{pp}}{2 \times \sqrt{2}}$$

$$= \frac{(22.0 - 2.0) \times (0.8)}{8}$$

$$= 2.0\text{W}.$$

The power thus calculated should be more than, or equal to, the required value of the power to be delivered, 2.0 Watts, in the present case.

3. Calculation of R_E

Since the minimum V_{CE} is assumed to be 2.0V, as in the above calculations, of which $V_{CE(sat)} = 1$V, the remaining voltage of 1.0 V will be across the resistance R_E.

This gives us the value of the resistance $R_E = 1.0 / 0.44$ A
$= 2.27$, say 2.2 Ohms.

4. Biasing Network

Since the stability is not specified, we may assume the value to be about 10 to 15. Hence, R_B = 10 to 15 times the value of the resistance R_E.

Let resistance R_B = 10 to 15 times R_E
= about 30 Ohms. (i)

But $$R_B = \frac{R_1 \times R_2}{R_1 + R_2}$$

and $$V_B = \frac{R_2}{R_1 + R_2} \times V_{CC}$$

From which, $R_1 = 11\,R_2$ (ii)

From Equations i and ii, we get

$$R_2 = 32.7 \text{ Ohms.}$$
$$R_1 = 360 \text{ Ohms.}$$

We shall select the nearest available values. Therefore,

R_2 = 33 Ohms.
R_1 = 370 Ohms.

To complete the design, we shall calculate the value of the capacitor, C_E, such that the reactance X_c = about 1/10th of the resistance R_E at around 100 Hz. (Assuming again that the lowest frequency of operation is about 100 Hz.)

Therefore, $$0.22 = \frac{1}{2 \cdot \pi \cdot 100\text{Hz} \cdot C}$$

Hence, $C = 1/(0.22 \times 2 \times \pi \times 100)$
= 7000 uF, which is a very large value.

As a matter of interest, let us calculate the efficiency of the designed amplifier.

When the output power = 2 Watts,
the *dc* power input = $V_{CC} \times I_{CQ}$
= 12×0.44A
= 5.28W.

This gives the efficiency

= $(2/5.28) \times 100$
= 37.8%

When no *ac* power is being delivered, i.e., at a time when *ac* wave-form passes through zero, the *dc* power input is still 5.28 Watts.

[**Note:** In power amplifier calculations, whenever a resistance value is calculated, it is imperative that its power rating be also calculated. [Refer to the next example].]

EXAMPLE 9.2

Design a class A power amplifier to deliver 6V rms to a load of 6 Ohms, using a transformer coupling. Assume that a supply of 18V is available. The resistance of the primary winding of the transformer also should be considered.

Solution

Refer to Fig 9.5.

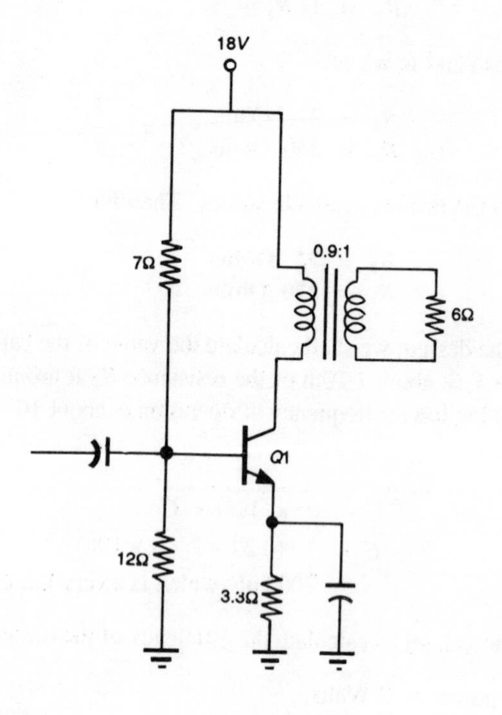

Fig. 9.5

1. Selection of the Transistor

The power output required is $= \dfrac{(V_{orms})^2}{R_L}$

$$= \frac{6 \times 6}{6} = 6\text{W}.$$

Assuming a transformer efficiency, η, of 90%, we have, the power required

$$= P_o / \eta$$
$$= 6 / .9 = 6.67\text{W}.$$

Therefore, we shall have to design the amplifier for 6.67W. Since the maximum efficiency of the transformer-coupled power amplifiers is 50%, the power dissipation capability of the transistor should be at least 3 to 4 times the power required to be developed.

For the transistor, therefore, the $P_{d(max)}$ should be

$$= 6.67\text{W} \times 3$$
$$= \text{about } 20 \text{ Watts.}$$

Let us select a transistor, ECN149, for the purpose.

This transistor has

$$P_{d(max)} = 30\text{W at } 25°\text{C}$$
$$I_{c(max)} = 4\text{A}$$
$$V_{CE(sat)} = 1\text{V}.$$

2. Choice of the Q-point

For transformer coupled amplifier, ideally, $V_{CEQ} = V_{CC}$. We shall assume the voltage across the resistance R_E as about 20% of the supply voltage, i.e.,

$$V_E = 0.2 \times 18\text{V} = 3.6\text{V}.$$

Since $V_{CE(sat)} = 1\text{V}$, and also to avoid the distortion near the saturation region, we shall take the quiscent point voltage

$$= V_{CEQ} = \text{about } 2/3\text{rd } V_{CC}, \text{ giving us}$$
$$V_{CEQ} = 12\text{V}.$$

The maximum swing available will be about 1V less ($V_{CE(sat)} = 1\text{V}$.) than the supply voltage of 12 Volts.

Hence for a power of 6.67W, we have,

$$6.67 = \frac{V_p}{\sqrt{2}} \times \frac{I_p}{\sqrt{2}}$$

giving $I_p = 1.21$ Amps.

Therefore, the Q-point is at 12V, 1.21A.

3. Choice of Resistance R_E

We have assumed voltage across the resistance R_E as equal to 3.6V, being about 20% of the supply voltage V_{CC}.

Therefore, $R_E = \frac{V_{RE}}{I_{CQ}} = \frac{3.6\text{V}}{1.12\text{A}}$

$= 2.975$ Ohms, say, 3.3 Ohms,

which is the nearest available standard value of the resistance. Let us recalculate the voltage across the resistance R_E.

The voltage $V_{RE} = 1.21\text{A} \times 3.3$ Ohms

$= 4\text{V}.$

The power dissipation of the resistance $R_E = (1.21\text{A})^2 \times 3.3$ Ohms

$= 4.82$ Watts.

Hence, we select the resistance $R_E = 3.3$ Ohms, 10 Watts.

4. Transformer

Secondary voltage = 6V (rms).

Let us calculate the primary voltage and, hence, the turns ratio. At Q-point, the *dc* voltage across the primary is

$$= V_{CC} - V_{CEQ} - (I_{CQ} \times R_E)$$
$$= 18 - 12 - (1.21 \times 3.3)$$
$$= 2\text{V}.$$

Giving *dc* resistance of the transformer $= 2\text{V} / 1.21\text{A}$

$= 1.66$ ohms.

The equivalent resistance on the primary of the transformer is equal to $R_{ac} - R_{\text{primary}}$

$$= \frac{V_p}{I_p} - R_{\text{primary}}$$
$$= (11 / 1.21) - 1.66$$
$$= 7.43 \text{ Ohms}.$$

The turns ratio $= \frac{6}{7.43}$

$= 0.9$

5. Choice of Resistance R_1 and Resistance R_2

Assuming $R_B = 10$ times R_E, for good stability, we have,

$$\frac{R_1 \times R_2}{R_1 + R_2} = 10\, R_E \qquad \text{(i)}$$

Also, since $V_E = 4V = V_B$,

we have $$4 = \frac{R_2}{R_1+R_2} \times 18$$

or $$R_1 = 3.5 \times R_2 \qquad \text{(ii)}$$

From Equations (i) and (ii) above, we have

$$R_1 = 42 \text{ Ohms.}$$

and $$R_2 = 12.5 \text{ Ohms.}$$

We select the nearest available values, as

$$\mathbf{R_1 = 47 \text{ Ohms.}}$$
$$\mathbf{R_2 = 12 \text{ Ohms.}}$$

The Power rating of these resistances are as under.

$$\text{For } R_1 = \frac{(V_{R1})^2}{R_1} = \frac{(18-4.6)^2}{47}$$

$$= 3.82\text{W}$$

$$\text{For } R_2 = \frac{(V_{B2})^2}{R_2} = \frac{(4.6)^2}{12}$$

$$= 1.76 \text{ W.}$$

Hence, we select, $\mathbf{R_1 = 47 \text{ Ohms, } 5 \text{ Watts.}}$

$\mathbf{R_2 = 12 \text{ Ohms, } 2.5 \text{ Watts.}}$

Let us calculate the maximum undistorted power available, which is equal $(V_p / \sqrt{2}) \times (I_p / \sqrt{2})$

$$= (11 / \sqrt{2}) \times (1.21 / \sqrt{2})$$

$= 6.66$ Watts, which is more than the required value.

The circuit efficiency

Useful power output = 6.66 Watts.

$$\text{Power input} = V_{CC} \times I_{CQ} + \frac{(V_{CC})^2}{R_1+R_2}$$

$$= 21.78 + 2.44 = 24.22 \text{ Watts.}$$

The circuit efficiency is, therefore,

$$= (6.66 / 24.2) \times 100$$
$$= 27.49\ \%$$

Likewise, let us calculate the transistor power dissipation when no signal is applied, which is

$$= V_{CEQ} \times I_{CQ}$$
$$= 12 \times 1.21$$
$$= 14.52 \text{ Watts.}$$

The power dissipation when the rated power is delivered

$$= 14.52 - 6.66$$
$$= 7.86 \text{ Watts.}$$

9.6 Class B Power Amplifier

The transistor connection in the circuit for class B amplifier remains nearly the same as that for the class A amplifier, except for the fact that the class B configuration needs the bias at zero or cut-off. But doing so only the positive half of the signal gets amplified, and hence such circuit can not be used for linear amplification applications. Class B amplifiers for such applications are, therefore, employed in push-pull configuration only. One such amplifier connection is shown in Fig. 9.6(a) and 9.6(b).

During the positive half of the signal, the transistor T_1 conducts and produces the amplification. The second transistor at the time remains off, since its base is driven beyond cut-off as shown in Fig. 9.6(a). During the other half of the signal waveform, the second transistor turns on, the first transistor having being forced off as shown in Fig. 9.6(b). Since the load is common, in the secondary of the transformer, we get across the load full wave amplified output voltage.

In class A power-amplifier, the power output can be increased by increase in the load resistance and to some extent, by increased signal input. But increase in the signal amplitude creates greater harmonic distortion due to close approach to cut-off. The advantages of the class B push-pull amplifier should be quite evident. It is possible to get higher output, since we are using two transistors, with each one of them providing greater excursion for the operating point, thus being able to accept higher level of signals. Since the transistors are biased at cut-off, they do not dissipate any power in the absence of signal voltage. This means that the stand-by power requirement for the amplifier has reduced to almost zero. This will increase the efficiency. However, due to reasons which we will study shortly, the distortion is higher, though even harmonic distortion cancels out. Yet another advantage is that due to the push-pull circuit, the saturation of the core is avoided.

Assuming a full length excursion, we have the maximum swing equal to V_{CC}. Hence the output power is equal to

$$P_{out} = \frac{V_{CC} \times I_p}{2} \tag{9.9}$$

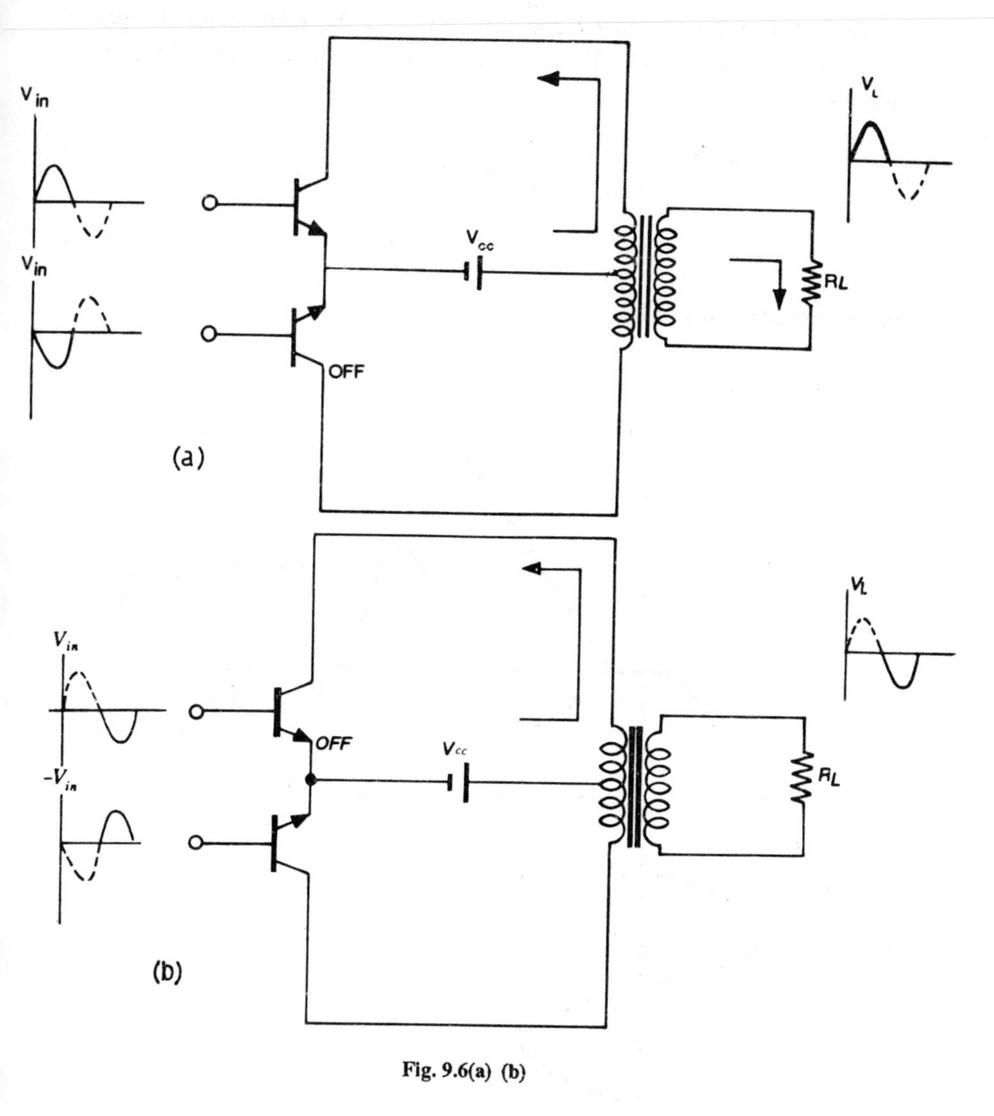

Fig. 9.6(a) (b)

The corresponding direct current in the transistor is the average value of half the sinusoidal current which flows through it.

i.e., $$I_{av} = I_p/\pi$$

Hence, the total power input = twice the power input to each of the transistors. Or

$$P_{in} = 2 \times \frac{V_{CC} \times I_p}{\pi} \tag{9.10}$$

The ratio of Equations 9.9 and 9.10 gives us the relation for the power efficiency of the amplifier. Thus, the maximum possible efficiency for class B amplifier is

$$\frac{P_{out}}{P_{in}} = \eta = \frac{\pi}{4}$$

$$= 78.5\% \tag{9.11}$$

9.7 Cross-over Distortion

Refer to Fig 9.7.

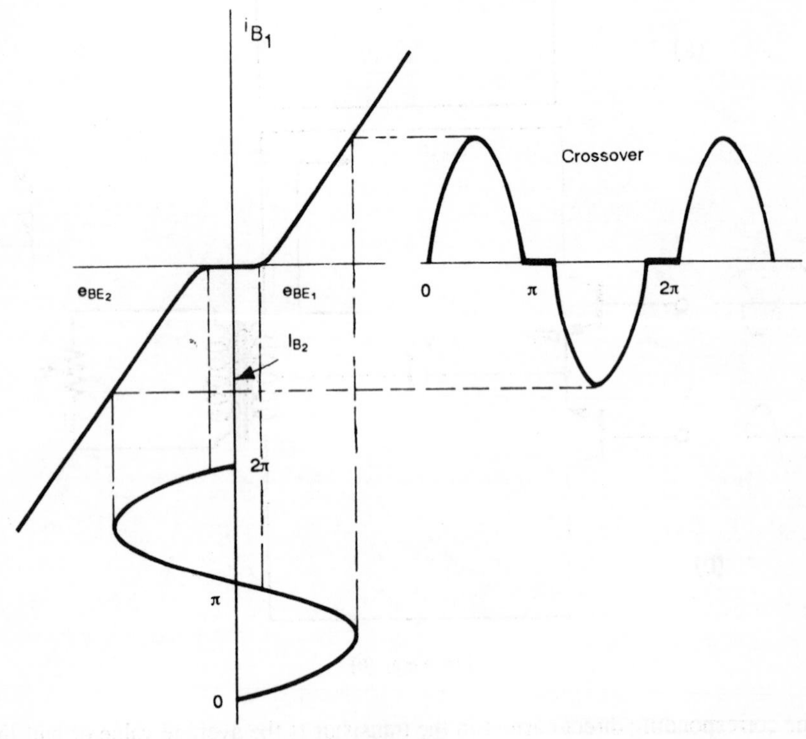

Fig. 9.7

Shown are the two assumed transfer curves for the respective transistors. If the Q point on the output transfer curve of transistor T_1 is moved down towards cut-off, a greater length of the transfer curve is available for the swing of the input signal. However, when the input signal goes through its negative cycle, the transistor gets cut-off, resulting into a considerable even-harmonic distortion. If you remember, we did not permit the Q point in class A power amplifier to wander near the saturation or cut-off region, even at a cost of reducing the power efficiency.

However, since we are using two transistors in a push-pull configuration, even harmonic distortion generated by each of the transistors is in different parts of the cycle. It can be shown mathematically that this configuration will cancel out the even harmonic generated.

What is important, however, is the transfer curve near the cut-off point. The base-emitter being a diode junction, it will not start conduction until the voltage V_{BE} is approximately 0.5 V. This means, therefore, that none of the transistors will be in conduction when the input signal is nearing zero, i.e., crossing from positive to negative, or from negative to positive value. The effect of this is shown in the output waveform in the figure. This can be avoided by bringing the two transfer curves together as shown in Fig. 9.8. The figure shows that before one transistor goes off, the other transistor is forced into conduction. This can very easily be done by providing a small positive bias to each of the transistors, as shown in Fig. 9.9. This keeps the transistors on the threshold of conduction. Effectively, the conduction angle does increase beyond 180°. However, since the bias provided is very small, this configuration can still be called class B.

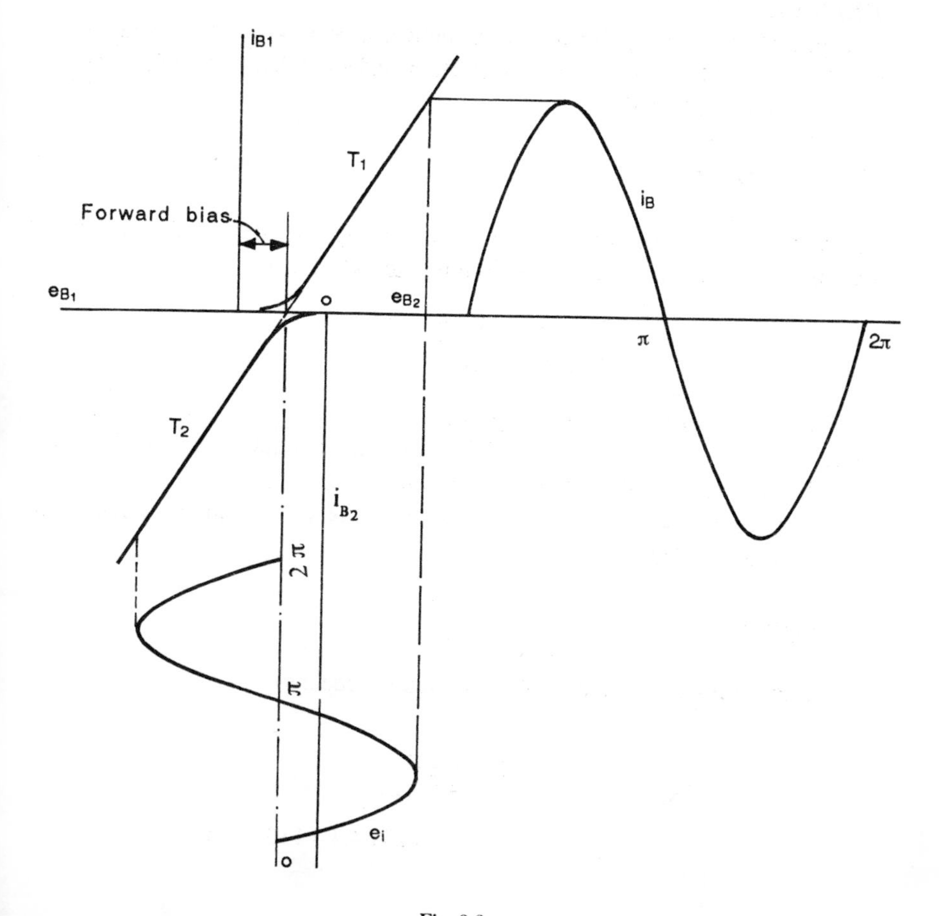

Fig. 9.8

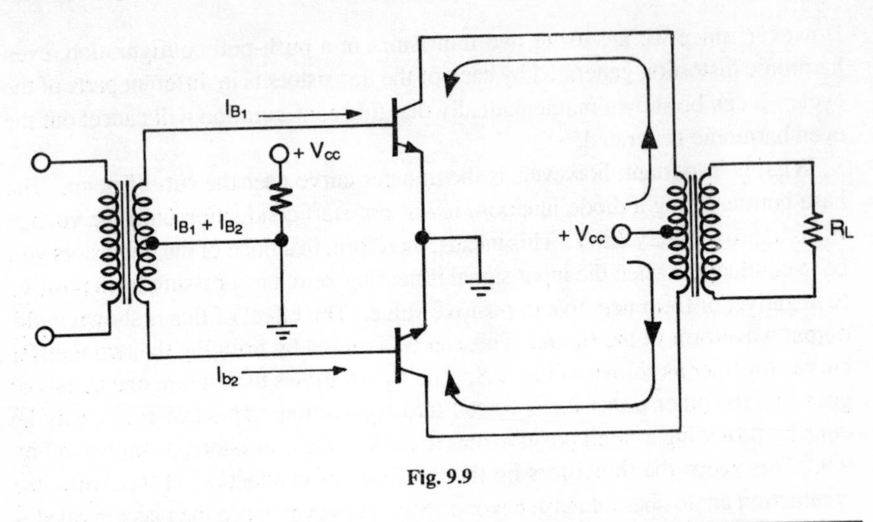

Fig. 9.9

EXAMPLE 9.4

Design a class B push-pull amplifier to deliver a peak-power of 2 Watts into a resistive load of 10 Ohms, with low distortion, and a stability factor of 8. Assume a supply voltage of 24 Volts.

Choice of Transistor

Since the transistor has to dissipate approximately one-fifth of the rated peak output power, we can choose ECN100 transistor which has

$$P_{dmax} = 5 \text{ Watts at } 25°\text{C},$$
$$I_{Cmax} = 0.7 \text{ A}.$$

The collector current $I_C = (2 \times P_o) / V_p$,
where P_o considered here is about 10% more than the required value of 2W, and V_p is the peak value of the voltage swing, which is equal to 24V.

This gives us $I_C = 2 \times 2.2 / 24 = 0.19$ A, which is well within the current limit of the transistor selected.

Transformer

The effective resistance in the collector circuit is equal to

$$R_c = \frac{V_{CC}^2}{2\,P_{o(max)}}$$

$$= 23 \times 23 / 4.4 = 110 \text{ ohms}.$$

Note: We have taken the supply voltage as 24 volts, whereas the actual swing is assumed, about one volt less, to be 23 V.

Since each transistor sees half the full-load for half the time, we have, collector-to-collector load as four times the one calculated above, i.e., $4 \times 110 = 440$ Ohms. This now gives us the transformer turns ratio, as

$$(440 / 10)^{0.5}$$

$$= 6.62 \text{ i.e., } 3.31 + 3.31{:}1$$

for the center-tapped output transformer.

Biasing Resistors

The emitter resistance, R_E, should have a low value, since it reduces the efficiency. Let R_E be equal to about one-tenth of R_C, equal to about 10 Ohms. Let us check for the power output. We have

$$P_{omax} = \frac{V_{CC}^2}{2\,(R_C + R_E)}$$

$$= (24 \times 24) / 2(110 + 10)$$
$$= 2.4 \text{ Watts.}$$

Out of this power, the useful power delivered to the load

$$= 2.4 \times \frac{R_C}{R_C + R_E}$$

$$= 2.4 \times 110 / (110 + 10)$$
$$= 2.2 \text{ Watts,}$$

which is slightly more than the power output required.

The values of the resistances R_1 and R_2 are so chosen as to create a bias voltage which is about 0.2 to 0.3 volts to avoid the cross-over distortion. The current drained off in this network may be limited to, say, 2 mA, which is very small current as compared to the operating current of the transistor, hence creating a power loss which is very small as compared to the useful power output.

With this assumption the total resistance = $V_{CC} / 2\text{mA}$

$$= 24 / 2\text{mA} = 12 \text{ k-Ohms} \qquad \text{(i)}$$

Hence, $R_1 + R_2 = 12$ K.

Also, since V_B is assumed to be about 0.2 V, we have,

$$\frac{R_2}{R_1+R_2} \times V_{CC} = 0.2 \text{ V}$$

giving $$R_1 = \text{about } 120\, R_2 \qquad \text{(ii)}$$

From Equations i and ii we have,

$$\mathbf{R_2 = 100 \text{ Ohms.}}$$
$$\mathbf{R_1 = 12 \text{ k-Ohms.}}$$

EXAMPLE 9.5

Design a class B AF power amplifier stage to give 10W *ac* power to a resistance load, 3 Ohms, which is transformer coupled. For the designed circuit, calculate the maximum undistorted power available and the corresponding input signal required, as also the circuit efficiency.

Solution

Refer to the Fig. 9.9

The power output to the load resistance of 3 Ohms is 10 W. If the transformer efficiency is assumed to be about 90%, the power delivered to the transformer primary should be

$$= 10\text{W}/0.9 = 11.11 \text{ W.}$$

The worst dissipation case for the transistor is about 20% of this value, giving P_{dmax} for the transistor as 0.2×11.11

$$= 2.22 \text{ Watts.}$$

Again we select a pair of transistors **ECN100** for the purpose. This transistor has

$$P_{dmax} = \mathbf{5 \text{ W at } 25°C,}$$
$$I_{Cmax} = \mathbf{0.7 \text{ A.,}}$$
$$V_{ceo\,max} = \mathbf{60V}$$

Since $V_{ceo\,max}$ is 60 volts, the supply voltage can not be more than 30V. To be on the safer side, we design the circuit to operate on 20V. With this voltage, the maximum swing possible is V_{CC} (voltage drops across R_E, transformer resistance and the transistor saturation voltage.)

Assuming the drop across R_E to be about 0.8 V, the drop across transformer primary (*dc*) resistance to be equal to about 0.5V, and the $V_{CE(sat)}$ as 0.8 Volts, we have maximum swing possible, as V_{p}, = 20 – (0.8 + 0.8 + 0.5) = 17.4 V.

The corresponding I_p can be calculated from the power that has to be delivered to the transformer primary. The power delivered to the transformer primary can be calculated as the product of the rms value of the voltage and the rms value of the current.

$$P'_L = (V_p / \sqrt{2}) \times (I_p / \sqrt{2})$$

Hence, the peak current $= I_p = (11.11 \times 2) / 17.4$ V.

$= 1.28$ A.

This calculated current, unfortunately, is greater than the maximum current that the transistor can handle. Therefore, the choice of the transistor will have to be remade. Let us select another transistor ECN 149. This transistor has the following parameters:

$$\mathbf{P_{D\,max} = 30 \text{ Watts at } 25°C}$$
$$\mathbf{I_{C\,max} = 4.0 \text{ A.}}$$
$$\mathbf{V_{ceo} = 40V}$$
$$\mathbf{h_{fe\,max} = 215.}$$

We shall have to calculate the current and the supply voltage again, since this transistor has different capacity. The voltage, V_{ceo} is = 40, hence the supply voltage can not be more than 20 volts. We shall select the supply voltage as equal to 15 V.

V_p, the maximum swing possible $= 15 - (0.8 + 0.8 + 0.5)$

$= 12.9$ volts.

It will be interesting to note the power consumption when no signal is present.

The voltage, V_{CQ}, $= 15 -$ (voltage drop across the resistance, R_E)

$= 15 - 0.8 = 14.2$ V.

The no-signal current can be assumed to be equal to 50 mA. The power per transistor, therefore, is 14.2 V $\times$ 50 mA

$= 0.71$ W.

The average power per half cycle for which each transistor works, can be calculated as

$$P_{av} = V_{CC} \times I_p$$
$$= 2.48 \text{ Watts.}$$

If we assume that the stability factor is about 10, then, we can say that

$$\frac{R_B}{R_E} = 10 \qquad \text{(i)}$$

Also, since
$$R_E = \frac{V_{RE}}{I_{CQ}}$$
$$= 0.8 / 50 \text{ mA}$$
$$= 16 \text{ Ohms.} \qquad \text{(ii)}$$

From Equations i and ii, we have
$$R_B = 160 \text{ Ohms.}$$

Again, if we assume that the voltages at the base and the emitter are equal, i.e., $V_B = V_E = 0.8$, as already calculated, we can write

$$R_B \times V_{CC} = \frac{R_2}{R_1 + R_2} \times V_{CC} = 0.8 \qquad \text{(iii)}$$

From which $R_1 = 17.75 \times R_2$

Hence $R_1 = 3000$ Ohms

and $R_2 = 169$ Ohms.

We can select the standard value for the above as

R_1 = 3300 Ohms.

and **R_2 = 175 Ohms.**

The effective load resistance, $R'_L = V_p / I_p$,
$$= 12.9 / 1.72$$
$$= 7.5 \text{ Ohms.}$$

But the collector-to-collector resistance is four times the above calculated value = 4 × 7.5 = 30 Ohms. Thus, the transformer should have center-tapped primary winding with the above resistances, and should have a power rating which is greater than 11.11 W. The power rating should be 15 Watts.

PART II

COMPUTER AIDED DESIGN OF ELECTRONIC CIRCUITS

1

DC Circuit Analysis

To solve a circuit for finding the current in a particular branch with the help of a computer needs a different type of approach. The computer will have to be fed the description of the circuit and, then, the method to solve this circuit.

1.1 Nodal Analysis

Refer to the circuit given in Figure 1.1.

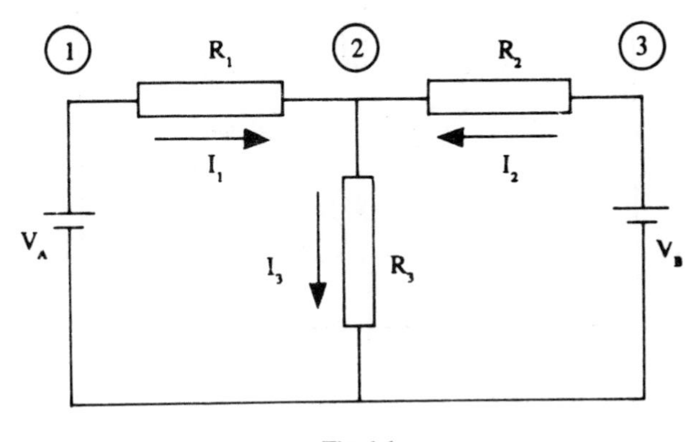

Fig. 1.1

It is required to find the current through the resistance R_3. We will solve this circuit by nodal analysis. The circled numbers in the circuit are the nodes, and the current directions are as shown in the figure.

The values of the voltages and the resistances are given. Now, if we find the voltage at a node 2, we will be able to calculate the current in R_3.

We can write for node 2, by KCL

$$\frac{V_1 - V_2}{R_1} + \frac{V_3 - V_2}{R_2} = \frac{V_2}{R_3}$$

where V_1, V_2 and V_3 are the voltages at nodes 1, 2, 3 respectively. Rearranging the above equation we get

$$V_2 \times \left(\frac{1}{R_1} + \frac{1}{R_2} + \frac{1}{R_3} \right) = -\frac{V_1}{R_1} = -\frac{V_3}{R_2} = 0 \qquad (1.1)$$

Since the inverse of a resistance is a conductance, we can write the equation 1.1 as

$$V_2 \times (G_1 + G_2 + G_3) - V_1 \times G_1 - V_3 \times G_2 = 0 \tag{1.2}$$

Let us try to derive an equation for a little more complex circuit. Refer to Fig. 1.2.

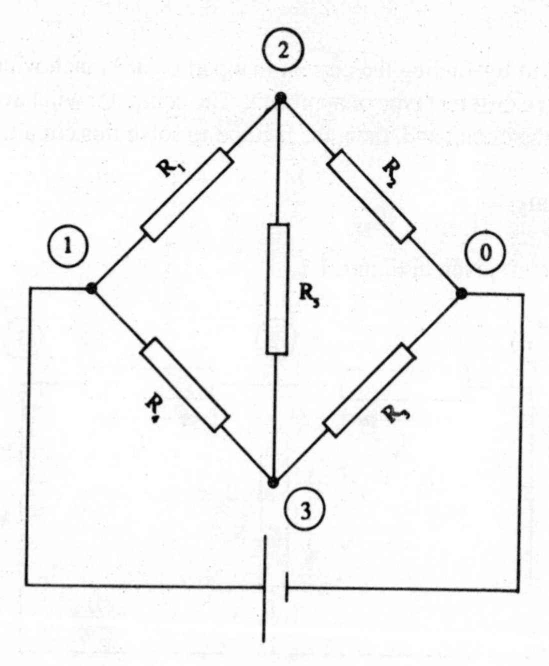

Fig. 1.2

Applying KCL at **node 2**, we have

$$\frac{V_1 - V_2}{R_1} - \frac{V_2 - V_0}{R_2} - \frac{V_2 - V_3}{R_5} = 0 \tag{1.3}$$

and at **node 3**, we have

$$\frac{V_1 - V_3}{R_4} - \frac{V_3 - V_0}{R_3} - \frac{V_2 - V_3}{R_5} = 0 \tag{1.4}$$

Rearranging, we get a set of equations as

$$V_2 \times \left(\frac{1}{R_1} + \frac{1}{R_2} + \frac{1}{R_5} \right) - \frac{V_1}{R_1} - \frac{V_0}{R_2} - \frac{V_3}{R_5} = 0 \tag{1.5}$$

and

$$V_3 \times \left(\frac{1}{R_3} + \frac{1}{R_4} + \frac{1}{R_5} \right) - \frac{V_1}{R_4} - \frac{V_0}{R_3} - \frac{V_2}{R_5} = 0 \tag{1.6}$$

or $$V_2 \times (G_1 + G_2 + G_5) - V_1 \times G_1 - V_0 \times G_2 - V_3 \times G_5 = 0 \tag{1.7}$$

and $$V_3 \times (G_3 + G_4 + G_5) - V_1 \times G_4 - V_0 \times G_3 - V_2 \times G_5 = 0 \tag{1.8}$$

Given the supply voltage **v** and all the branch resistances, we should be able to calculate the current in R_5 by Equations 1.7 and 1.8, keeping in mind that V_1, the voltage at node 1 is equal to the supply voltage V, and the voltage at node 0 = 0, being a reference node in the figure.

From Equation 1.2 for the circuit of Fig. 1.1, and Equations 1.7 and 1.8 for the circuit of Fig. 1.2, a pattern seems to be emerging. We can make the following statements :

For any node in the circuit, we can write KCL equations for that node by (i) taking the voltage at that node and multiplying it with a sum of all the conductances at that node, and subtracting therefrom, all products of the node voltage and the connected conductance between them and the node under analysis. For the explanation of the above, take Equation 1.7. Here, node 2 is under analysis. Hence we get $V_2 \times (G_1 + G_2 + G_5)$ as per (i) above, subtracting therefrom, $V_1 \times G_1$, $V_0 \times G_2$, $V_3 \times G_5$, where V_1, V_0 and V_3 are the voltages of the other three nodes connected to node 2, through G_1, G_2 and G_5, respectively. The reader can check Equations 1.2 and 1.8 on similar basis and verify the statement made above.

From this equation it should be clear that we can find node voltages by the following matrix operations

$$[V] \times [G] = [K] \tag{1.9}$$

where $[V]$ is the node voltage matrix, $[G]$ is the conductance matrix and $[K]$ is the constant vector

1.2 Solution by Computer

The above problems can be solved by

$$[V] = [G]^{-1} \times [K]$$

This involves taking an inverse of the matrix G which can be constructed by invoking a procedure for the purpose in the main computer program.
The method employed could be the Gaussian reduction method. This can also be achieved by determinant operation as

$$G^{-1} = \frac{\text{Adj}\,(G)}{|G|} \tag{1.10}$$

When it comes to a solution by the computer, however, many difficulties will arise.

The solution by matrix operation is possible provided matrices are available. The question is how is the computer going to generate the matrices. Let us ponder on this point at some length.

There are softwares available, which need topological information from which they can map the circuit in the computer memory and, after generating the matrices, can invoke several connected procedures to analyse the circuit. One such software is SPICE. This software requires the user to create a file, giving number of nodes and other topological information in a manner specified by the software. Such a file can be created by, usually, the editor of the system on which one is working. The SPICE, then, generates another file which contains the analysed data.

Here, similar program is written in **BASIC**, in an inter-active way which will prompt the user for feeding the data. The program is discussed in parts hereafter. We will discuss the method for the circuit shown in Fig. 1.1, where all the nodes are to be specified. The voltage sources V_A and V_B will create problems. In most of the favourable cases, we neglect the internal resistance of the source (assuming that the voltage source is ideal) which, otherwise, would present a node within the source, making analysis a little more complex and time consuming. However, taking voltage source as ideal does present its own problem, since the nodal analysis is based on the fact that the current meeting at a point is equal to zero. The circuit will have to be redrawn, the voltage source having been replaced by the equivalent current source.
Refer to Fig. 1.3.

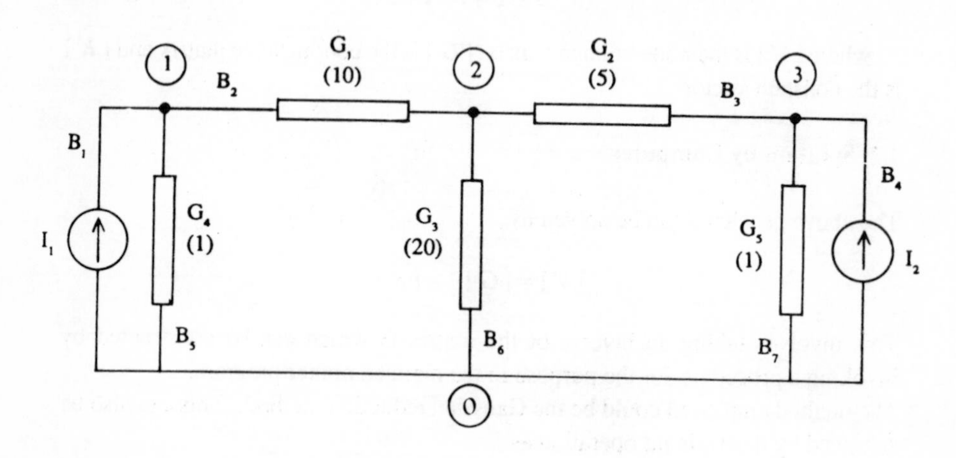

Fig. 1.3

R_4 and R_5 are the internal resistances of the sources.
The set of equations is

$$V_1 \times (G_1 + G_4) - V_2 \times G_1 = I_1 \tag{1.11}$$

$$-V_1 \times G_1 + V_2 \times (G_1 + G_2 + G_3) - V_3 \times G_2 = 0 \tag{1.12}$$

$$-V_2 \times G_2 + V_3 \times (G_2 + G_5) = I_2 \tag{1.13}$$

This generates the three matrices as

$$\begin{bmatrix} \mathbf{V_1} \\ \mathbf{V_2} \\ \mathbf{V_3} \end{bmatrix} \begin{bmatrix} \mathbf{G_1 + G_4} & \mathbf{-G_1} & 0 \\ \mathbf{-G_1} & \mathbf{G_1 + G_2 + G_3} & \mathbf{-G_3} \\ 0 & \mathbf{-G_2} & \mathbf{G_2 + G_5} \end{bmatrix} = \begin{bmatrix} \mathbf{I_1} \\ 0 \\ \mathbf{I_2} \end{bmatrix}$$

The number of entries in V matrix should be equal to the number of nodes in the circuit (node 0 not to be counted). The G matrix should be a square matrix ($N \times N$), where N is the number of nodes. Let us create a file of data and write the information available in a tabular form.

In Table 1.1, numbers 1 to 7 denote different branches. P and N are the +ve and –ve ends of the branches, or indicate the start and end-node position of the branch.

Table 1.1

Br. No.	*P*	*N*	*Type*		*Value*
1	1	0	I		5
2	1	2	G	(G_1)	10
3	2	3	G	(G_2)	5
4	3	0	I		10
5	1	0	G	(G_4)	1
6	2	0	G	(G_3)	20
7	3	0	G	(G_5)	1

Now let us see the way the conductance matrix can be created.

Step 1 We initialise the G matrix and matrix I by making all the elements equal to zero, i.e.

$$G_{ij} = 0 \text{ for all } i\text{'s and all } j\text{'s.}$$

and

$$I_i = 0 \text{ for all } i\text{'s}$$

$$\mathbf{G} = \begin{bmatrix} 0 & 0 & 0 \\ 0 & 0 & 0 \\ 0 & 0 & 0 \end{bmatrix} \text{ and } \mathbf{I} = \begin{bmatrix} 0 \\ 0 \\ 0 \end{bmatrix}$$

Step 2 Now we fill up the matrix.

The nodes are P and N. Hence, we start scanning Table 1.1, then

(a) $G(P, P) = G(P, P)$ + (corresponding G)

where P should vary from 1 to total number of nodes, $S = 3$, in our case. We scan the table for P, starting from branch 1. We skip the branches which contain node zero in either P or N column, or the ones which do not contain conductances.

Branch 1 contains $N = 0$ and the type is I, hence we skip it

Branch 2 contains $P = 1$ and the type is G, hence we write

$$G_{11} = G_{11} + G_1$$

Branch 3 has $P = 2, N <> 0$, and conductance G_3 is present

Hence, $$G_{22} = G_{22} + G_2$$

We skip branches 4 to 7 since node N = 0 in each case.

The matrix is now modified and looks as follows.

$$\begin{bmatrix} G_1 & 0 & 0 \\ 0 & G_2 & 0 \\ 0 & 0 & 0 \end{bmatrix}$$

(b) We repeat the above procedure for

$G(N, N) = G(N, N)$ + (corresponding G),

with node N varying from 1 to N (= 3 in our case).

The matrix will look like

$$\begin{bmatrix} G_1 & 0 & 0 \\ 0 & G_1 + G_2 & 0 \\ 0 & 0 & G_2 \end{bmatrix}$$

(c) $G(N, P) = G(N, P)$ – (corresponding G)

$G(N, P)$ for value of N and P given in each branch (again with a condition that either P or N should not be zero and the type should be conductance).

Branch 2 and branch 3 yield two more elements (all other branches will be skipped, because of the above condition) After this pass, this matrix is,

$$\begin{bmatrix} G_1 & 0 & 0 \\ -G_1 & G_1 + G_2 & 0 \\ 0 & -G_2 & G_2 \end{bmatrix}$$

(d) We repeat the procedure for

$$G(P, N) = G(P, N) - (\text{corresponding } G)$$

giving the matrix as :

$$\begin{bmatrix} G_1 & -G_1 & 0 \\ -G_1 & G_2 + G_1 & -G_2 \\ 0 & -G_2 & G_2 \end{bmatrix}$$

Step 3 Again we scan the table to find values between the nodes and the reference node. The node can be either P or N,

$$G(K, K) = G(K, K) + (\text{corresponding } G)$$

Branch Nos 5, 6 and 7 have conductance and node N equal to zero. Hence these branches will contribute elements to the matrix

$$\begin{bmatrix} G_1 + G_4 & -G_1 & 0 \\ -G_1 & G_1 + G_2 + G_3 & -G_2 \\ 0 & -G_2 & G_5 + G_2 \end{bmatrix}$$

The creation of conductance matrix is now complete. On similar lines the I matrix can also be created by taking

$$I(P) = I(P) - \text{value of current } I, \text{ if present}$$
$$I(N) = I(N) - \text{value of } I, \text{ if present}$$

Hence first element, I_1 = value of current I_1 (from the table), and element I_3 will have current I_2.

The I vector should look like

$$\begin{bmatrix} I_1 \\ 0 \\ I_2 \end{bmatrix}$$

Thus,

$$\begin{bmatrix} V_1 \\ V_2 \\ V_3 \end{bmatrix} \begin{bmatrix} G_1 + G_4 & -G_1 & 0 \\ -G_1 & G_1 + G_2 + G_3 & -G_2 \\ 0 & -G_2 & G_5 + G_2 \end{bmatrix} = \begin{bmatrix} I_1 \\ 0 \\ I_2 \end{bmatrix}$$

Now we need to find the current in each branch

Again we refer to Table 1.1. Voltages at all the nodes are now available. Hence, we can now find current in each branch and its direction by following algorithm.

We find CU(branch) = [$V(P) - V(N)$] / conductance, provided Type = G
If the current is positive, it is flowing from node (P) to node (N).

If the current is negative, the flow is from node (N) to node (P).

In Fig. 1.3, only two current sources are shown. Both these sources, conveniently, are shown from one of the nodes to the reference node. There could be two variations to this arrangement. The first variation could be the existence of a current source between any two nodes. i.e., one of the nodes need not be a reference node. The other variation could be that such a current source could also be a dependent current source, the magnitude depending on some other voltage (Voltage Controlled Current Source) or some other current (Current Controlled Current Source) in the network under analysis. The procedure for generating the matrices will remain, essentially the same. The program will need some minor modifications.

1.3 Computer Program for Nodal Analysis

The **PASCAL** (turbo) version of the program is given in Computer Programs P2., P3 and P4. Program P2 is interactive, while P3 and P4 are based on records and file operations. The programs use Gauss Elimination method for solution of the set of the simultaneous equations.

The *ac* network analysis should also proceed in exactly the same way, the only difference being that the conductance matrix will now be called the admittance matrix. The laplace notations should provide easier method of solution compared to the complex method of representing the admittance.

Though Gaussian reduction method is assumed here because of its popularity, any other suitable method may be used. Depending on the network under analysis, the lower and upper triangle factorisation (LU) may also be used, as also the sparsing properties can be made use of for simplification, or for getting better precision. The details of these methods are not discussed here, nor their relative merits, as it is a matter of numerical analysis suited to computer application and sufficient knowledge of the same is assumed.

Before we employ any kind of nodal analysis to the electronic circuits, it becomes imperative to familiarise with the models of the various devices. In the next chapter, we shall discuss some of the device models.

PROGRAM 1

```
10  PRINT "HOW MANY BRANCHES AND NODES?"
20  INPUT NO, NODES
30  DIM BR(NO), P(NO), N(NO)
40  PRINT "THE INFORMATION BR.NO. P N TYPE VALUE"
50  PRINT "SEPARATE THE VALUES BY COMMAS"
60  FOR J = 1 TO NO
70  READ BR(J), P(J), N(J), TYPE$(J), VALUE(J)
80  NEXT J
90  PRINT "------------------------------------------------------------"
100 PRINT
110 PRINT "   BR. NO.   P        N        TYPE     VALUE   "
120 PRINT
130 PRINT "------------------------------------------------------------"
140 FOR J = 1 TO NO
150 PRINT TAB(10);BR(J);TAB(19);P(J);TAB(27);N(J);TAB(37);TYPE$(J);
    TAB(48)VALUE(J)
160 NEXT J
165 PRINT
166 PRINT "------------------------------------------------------------"
170 PRINT "ARE THE VALUES CORRECT?"
180 INPUT T$
190 IF T$ = "N" OR T$ = "n" GOTO 10
200 PRINT "NOW THE CONDUCTANCE MATRIX WILL BE PREPARED"
210 REM INITIALISATION OF THE G AND I MATRICES
220 FOR I = 0 TO NODES
230 I(I) = 0
240 FOR J = 0 TO NODES
250 G(I,J) = 0
260 NEXT J
270 NEXT I
280 REM "*******************************************"
290 PRINT
300 FOR J = 1 TO NO
310 IF (P(J) <> 0 AND N(J) <>0 AND TYPE$(J)<>"i") GOTO 340
320 GOTO 380
340 G(P(J),P(J)) = G(P(J),P(J)) + VALUE(J)
350 G(N(J),N(J)) = G(N(J),N(J)) + VALUE(J)
360 G(P(J),N(J)) = G(P(J),N(J)) – VALUE(J)
370 G(N(J),P(J)) = G(N(J),P(J)) – VALUE(J)
380 NEXT J
390 FOR J = 1 TO NO
```

```
400 IF P(J) = 0 AND N(J) <> 0 AND TYPE$(J) = "G" GOTO 420
410 GOTO 430
420 G(N(J),N(J)) = G(N(J),N(J)) + VALUE(J)
430 NEXT J
440 FOR J = 1 TO NO
450 IF N(J) = 0 AND P(J) <> 0 AND TYPE$(J) = "G" GOTO 470
460 GOTO 480
470 G(P(J),P(J)) = G(P(J),P(J)) + VALUE(J)
480 NEXT J
500 REM PRINT THE MATRIX
510 FOR I = 1 TO NODES
520 FOR J = 1 TO NODES
530 PRINT TAB(10*J); G(I,J);
540 NEXT J
550 PRINT
560 NEXT I
580 DATA 1,1,0,I,5
590 DATA 2,1,2,G,10
600 DATA 3,2,3,G,5
610 DATA 4,3,0,I,10
620 DATA 5,1,0,G,1
630 DATA 6,2,0,G,20
640 DATA 7,3,0,G,1
```

PROGRAM 2

```
PROGRAM NODALANALYSIS(INPUT,OUTPUT);
    TYPE MAT = ARRAY [1..20,1..21] OF REAL;
VAR
        P,N:ARRAY [1..50] OF INTEGER;
          E,VALUE:ARRAY [1..50] OF REAL;
        VOLT,CRNT:ARRAY [0..50] OF REAL;
       ETYPE:ARRAY[1..50] OF CHAR;
           G:MAT;
           NOBR,NODES,J,K;INTEGER;

    (PROCEDURES SECTION)

     PROCEDURE READDATA;
     VAR  J:INTEGER;
        BEGIN



             FOR J:=1 TO NOBR DO
        BEGIN
             WRITELN('ENTER P,N,ETYPE,VALUE');
      READ(P[J]);READ(N[J]);READ(ETYPE[J]);READ(VALUE[J]);WRITELN;
         END;
       END;

PROCEDURE MATFORM;
VAR C,K,J:INTEGER;
BEGIN
    FOR J:=1 TO NODES DO
                        FOR K:=1 TO NODES+1 DO G[J,K]:=0;
C:=1;
WHILE C<=NOBR DO
BEGIN
    IF ETYPE[C]<>'I' THEN
      BEGIN
            IF (P[C])<>0)AND(N[C]<>0) THEN
                  BEGIN
                        G[P[C],P[C]]: = G[P[C],P[C]] + 1/VALUE[C];

                        G[N[C],N[C]]: = G[N[C],N[C]] + 1/VALUE[C];

                        G[P[C],N[C]]: = G[P[C],N[C]] – 1/VALUE[C];

                        G[N[C],P[C]]: = G[N[C],P[C]] – 1/VALUE[C];

                END
           ELSE
                IF P[C]<>0 THEN
                         G[P[C],P[C]]:=G[P[C],P[C]]+1/VALUE[C]
```

```
                                        ELSE
                                        G[N[C],N[C]]:=G[N[C],N[C]]+1/VALUE[C];
        END
      ELSE
          BEGIN
              IF P[C]<>0 THEN G[P[C],NODES+1]:=G[P[C],NODES+1]+VALUE[C];
              IF N[C]<>0 THEN G[N[C],NODES+1]:=G[N[C],NODES+1]-VALUE[C];
            END;
       C:=C+1;
       END;
    END;
 PROCEDURE JORDAN(VAR A:MAT);
 CONST Z=20;
 VAR
   C:REAL;
   I,J,K:INTEGER;
 BEGIN
    FOR I:=1 TO NODES DO
   BEGIN
        C:=A[I,I];
        FOR J:=1 TO NODES+1 DO A[I,J]:=A[I,J]/C;
        FOR K:=1 TO NODES DO
        BEGIN
            IF K<>1 THEN
                   BEGIN
                        C:=A[K,I];
                        FOR J:=I NODES+1 DO A[K,J]:=A[K,J]-(A[I,J]*C);
                   END;
        END;
    END;
           WRITELN('VOLTAGE AT EACH NODE IS . . . ');WRITELN;WRITELN;
    FOR I:=1 TO NODES DO
    BEGIN
WRITELN(A[I,NODES+1]:15:3);
    END;READLN;
END;
    PROCEDURE FINDCURRENT;
    VAR J,K:INTEGER;
    BEGIN
        FOR J:=1 TO NOBR DO
        BEGIN
            IF ETYPE[J]='G' THEN
                                    CRNT[J]:=(VOLT[P[J]]-VOLT[N[J]])/VALUE[J]
                              ELSE
                                    IF [P[J]=0 THEN CRNT[J]:=-VALUE[J]
                                                   ELSE CRNT[J]:=VALUE[J];
     END;
END;
```

```
( MAIN PROGRAM )
BEGIN
    WRITELN(' ENTER NO. OF BRANCHES & NO. OF NODES');
    READLN(NOBR,NODES);WRITELN;WRITELN;
    WRITELN(' DATA ENTRY STARTS NOW !!!');WRITELN;WRITELN;
    READDATA;
    MATFORM;
    JORDAN(G);
    VOLT[0]:=0;
    FOR J:=1 TO NODES DO VOLT[J]:=G(J,NODES+1];
    FINDCURRENT;
    WRITELN('CURRENT IN EACH BRANCH IS . .');
    FOR J:=1 TO NOBR DO WRITELN(' I[',J,'] = ',CRNT[J]*1000:10:4,'mA');
END.
```

PROGRAM 3

```
program nodal(input,output,nodeinfo),
type node=record
      vtype:char;
      p,n:integer;
      value:real;
end;
var g:array[1..25,1..25] of real;
   v,i:array[1..25] of real;
   branches,m,nodes,p,n,a,q:integer;
   c1,big,temp,value:real;
   node1:node;
   nodeinfo:file of node;
procedure valuefile;
begin
    writeln('how many branches in the ckt.');
    readln(branches);
    rewrite(nodeinfo);
    with node1 do
      begin
          for m:=1 to branches do
            begin
               writeln('i/p the component type as conductance(g) or current)
               readln(vtype);
               writeln('i/p the 2 nodal endpts of the component as p + & n+)
               readln(p,n);
               writeln('i/p the value of the component',m);
               readln(value);
               write(nodeinfo,node1);
            end;
      end;
end;
begin(*main program*)
    assign(nodeinfo,'a:nodeinfo.pas');
    writeln('if you are interested in new sets of values type 1');
    readln(a);
    writeln('how many branches in the ckt');
    readln(branches);
    writeln('how many nodes in the ckt.');
    readln(nodes);
    writeln;
    writeln;
    for p:=1 to nodes do
    begin
      i[p]:=0;
      for n:=1 to nodes do
```

```
    g[p,n]:=0;
  end;
    if a=1 then valuefile;
    reset(nodeinfo);
    for m:=1 to branches do
    begin
      read(nodeinfo,node1);
      p:=node1.p;
      n:=node1.n;
      value:=node1.value;
      if node1.vtype='g' then
                    begin
                      if(p<>0) and (n<>0) then
                      begin
                        g[p,p]:=g[p,p]+value;
                        g[n,n]:=g[n,n]+value;
                        g[p,n]:=g[p,n]-value;
                        g[n,p]:=g[n,p]-value;
                      end
                      else
                        begin
                          if p<>0 then g[p,p]:=g[p,p]+value;
                          if n<>0 then g[n,n]:=g[n,n]+value;
                        end;
                      end
                      end i[p]:=value;
        end;
                      writeln('initial conductance matrix is:=');
                      writeln;
                      writeln;
                      for p:=  1 to nodes do
                        begin
                          for n:=1 to nodes do
                            begin
                              write(g[p,n]:8:3,' ');
                              end;
                            writeln;
                        end;
                        write('the current matrix is...');
                        writeln;
                        writeln;
                        for p:=1 to nodes do
                          writeln(i[p]:12:1);
                          writeln;
(* gauss elimination *)
for p:=1 to nodes do
    begin
      n:= nodes + 1;
```

```
      g[p,n]:=i[p];
    end;
  for p:=1 to nodes do
    begin
      big:=g[p,p];
      for n:=p to nodes do
      begin
        if g[n,p]>big then
                                begin
                                  big:=g[n,p];
                                  for q:=1 to (nodes + 1) do
                                begin
                                   temp:=g[n,q];
                                   g[n,q]:=g[p,q];
                                   g[p,q]:=temp;
                                end;
                            end;
    end;
   for n:=(p+1) to nodes do
     begin
       while g[n,p]<>0 do
        begin
          c1:=g[p,p]/g[n,p];
          for q:=1 to (nodes+1) do
           g[n,q]:=g[p,q]–c1*g[n,q];
        end;
     end;
   end;
   writeln;
   writeln;
   writeln('The final conductance matrix:                    The current matrix:
   writeln;
   writeln;
   for p:= 1 to nodes do
   begin
    for n:= 1 to nodes+1 do
   begin
    If n=nodes+1 then write('                                        ',g[p,n]:8:3)
    else
       write(g[p,n]:8:3,'          ');
   end;
   writeln;
   end;
 v[nodes]:=(g[nodes,nodes+1])/(g[nodes,nodes]);
 c1:=0;
 for n:=(nodes–1) downto 1 do
   begin
     for q:=(n+1) to nodes do
```

```
    c1:=c1+v[q]*g[n,q];
    v[n]:=(g[n,nodes+1]-c1)/g[n,n];
end;
writeln;
writeln;('CONDUCTANCE MATRIX × VOLTAGE MATRIX=CURRENT MATRIX');
writeln;
writeln;(' The Nodal Voltage Matrix is as shown below  ');
writeln;
writeln;
for n:=1 to nodes do
 writeln('v[',n,']='v[n]:8:3);
readln;
end.
```

PROGRAM 4

```
PROGRAM cir(input,output,datafile);

VAR nodes,p,n,q,e            :  integer;
     btype                   :  char;
     datafile                :  text;
      c1,cvalue,big,temp     :  real;
      G                      :  ARRAY[1..25,1..25] OF real;
      V,I                    :  ARRAY[1..25] OF real;
      filevar                :  string[14];
      dataline               :  string[30];
      p1,n1,btype1,cvalue1   :  string[2];

BEGIN
 clrscr;
 write('ENTER THE NAME OF THE DATAFILE : ') ;
 readln(filevar) ;
 ASSIGN(datafile,filevar) ;
 write('ENTER THE NO. OF NODES IN THE CIRCUIT : ') ;
 readln(nodes);
 RESET(datafile) ;
  WHILE NOT (eof(datafile)) DO
   BEGIN
    readln(datafile,dataline) ;
    p1 := copy(dataline,5,1) ;
    val(p1,p,e) ;
    n1 := copy(dataline,7,1) ;
    val(n1,n,e) ;
      btype := copy(dataline,9,1) ;
      cvalue1 := copy(dataline,11,2);
      val (cvalue1,cvalue,e);
    IF btype = 'G' THEN
     BEGIN
      IF (p <> 0) AND (n <> 0) THEN
       BEGIN
        G[p,p] := G[p,p] + cvalue;
        G[n,n] := G[n,n] + cvalue;
        G[p,n] := G[p,n] - cvalue;
        G[n,p] := G[n,p] - cvalue;
       END
      ELSE
       BEGIN
        IF p = 0 THEN G[p,p] := G[p,p] + cvalue;
        IF n = 0 THEN G[n,n] := G[n,n] + cvalue;
       END
     END
```

```
    ELSE
      I[p] := cvalue;
    END;
```

Followed by gauss-elimination as in previous programs.
A text file called data file is created for type and value

2

Device Modelling

Any computer program will recognise and permit only certain well defined blocks within its analysis. Certain limited flexibility may be provided, making the program look a bit more 'intelligent'. However, there is always a danger of the program straying off the natural path of the process.

A better way has to be found to make the program more flexible. A transistor, for instance, behaves differently when working at two widely different frequencies. The program can not be so rigid as not to take this variation into account, nor can it be made to understand this on its own.

It is, therefore, imperative that the devices be presented to the circuit analysis program in such a way that the program does not have to be imbibed with natural intelligence, were it really possible.

We shall, therefore, try to present to the analytical program, a device model which becomes acceptable to it. The devices that can be considered are diodes, and transistors, both BJT and FET, etc. for which equivalent circuits will have to be developed. Simple devices are components like resistance, inductance, capacitance, and the voltage and the current sources, both dependent and independent.

A simple device like resistance also behaves differently at widely different frequencies. A pure resistance at low frequencies will exhibit a small angle of lag at high frequencies, i.e., at several hundred Mega Hertz, indicating a presence of inductance. It may also produce an angle which, as mentioned above, goes on increasing with frequency upto several hundred MHz and thereafter goes on decreasing with further increase in the frequency, indicating an increasing and dominating presence of capacitance which produces nearly resonance condition. At extremely high frequencies, where the wavelength becomes smaller than the physical length of a resistor, the behaviour will be like that of a transmission line.

Let us discuss some of the device models.

2.1 Junction Diode

Since we are interested in some moderately low frequencies, we will restrict ourselves to generally accepted models. The model for the junction diode is shown in Figure 2.1.

Shown in the figure is the symbolic notation and the model useful for dynamic analysis. The model consists of essentially, five parameters. These are as under:

1. R_d **:** This is a non-linear resistor, and represents the diode junction. The current, I_d's dependence on voltage, V_d is given by the following relation:

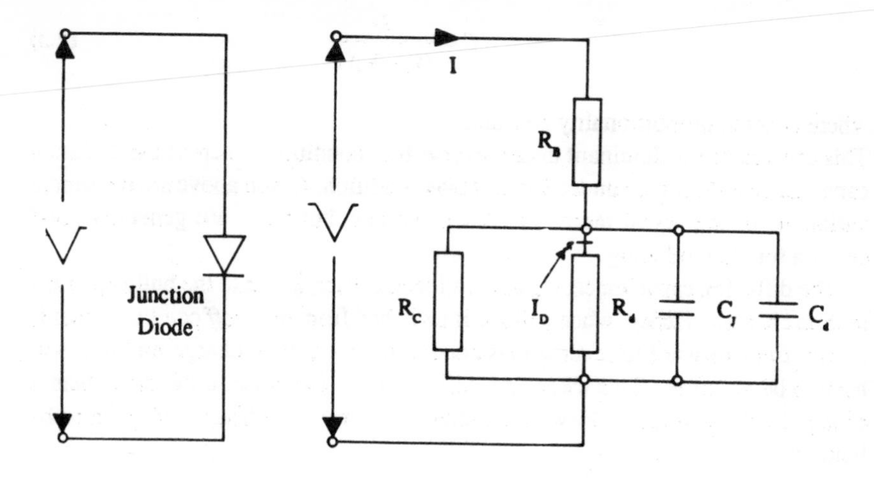

Fig. 2.1

$$I_d = I_s \times \left\{ \exp\left(\frac{q\, V_d}{k\, M\, T} \right) - 1 \right\} \tag{2.1}$$

where I_s = Diode saturation current. (This current is of the order of 10_A^{-12} to a high of about 10_A^{-6}, for silicon diodes and of the order of 10_A^{-8} to a high of about tens of micro-amps.

q = Electron charge = 1.6×10^{-19}

k = Boltzmann's constant = 1.38×10^{-23} J/°C.

M = Essentially is a correction factor and its value lies between 1 and 2.5.

T = Junction absolute temperature in degrees Kelvin.

2. R_B : The semiconductor bulk resistance and the contact resistance. (About 100 Ohms.)

3. R_C : The reverse resistance of the junction. (More than 1 Mohm).

4. C_d : The junction diffusion capacitance, or charge storage capacitance. This capacitance is given by

$$C_d = \frac{q}{2 - M\, k\, T\, f}(I_d + I_s) \tag{2.2}$$

where f is the intrinsic diode frequency.

It should be noted that the capacitance, C_d, is dependent on the current I_d, which in turn is dependent on the voltage V_d.

This capacitance is a serious limitation in high frequency circuit operation.

5. C_j : The junction transition capacitance is given by

$$C_j = \frac{D}{(V_t - V_d)\,t^n} \tag{2.3}$$

where D is the proportionality constant
This capacitance is dominant under reverse-bias condition, whereas the diffusion capacitance is dominant under forward bias condition. Given above are the simple variations of the actual more complex equations, but these are generally good enough for the modelling.

The diffusion capacitance, C_d, as mentioned earlier, is due to the build up of the mobile charges. Hence, when a diode is switched from *on* to *off* condition, that is to say, from forward to reverse bias condition, this mobile charge build-up will have to be removed. This takes up a definite time. Likewise, if the diode bias is changed from reverse to forward condition, this mobile charge will have to be built up.

The diode model shown above is obviously an *ac* model. If we are interested only in true *dc* circuits, the capacitances can be removed from the figure. The only thing we will have to worry about is the temperature variation effect.

2. Temperature Dependence

It has already been seen that the diode current is given by

$$I_d = I_s\,[\,\exp\,(V_d\,/\,n\,V_T) - 1] \tag{2.4}$$

where V_T = $T / 11600$
= 26 mV at temperature T = 300°K.

Hence, the characteristics indicating the variation of diode current, I_d, with reference to the bias voltage, V_d, is as shown in Fig. 2.2

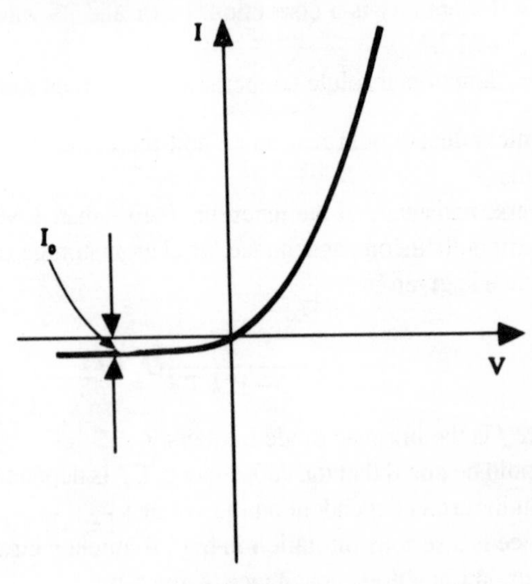

Fig. 2.2

It should be noted that the reverse saturation current, I_s, is given by the equation

$$I_s = k\,T^m \exp\left(-V_{go}/n\,V_T\right) \tag{2.5}$$

For silicon, $n = 2$, $m = 1.5$ and $V_{go} = 1.21$ V, for small current operations. Experimentally it is found that the reverse saturation current, I_s, increases at the rate of approximately 7%/°C, or approximately doubles for every 10°C rise in the temperature. It is also observed that after the diode (silicon) starts conduction (after about 0.6 V forward bias), the voltage, V_d, changes with the temperature at a rate of about −2.5 mV/°C.

THE RESISTANCE OF THE DIODE

The resistance of the diode as shown in the model also greatly varies, depending on the point of operation. The static resistance, following Ohm's law, is defined at a particular point of operation. For instance, if the diode voltage drop is of the order of, say, 0.7 volts at the point of operation, and the current at the point is, say, 10 mA, then the resistance can be calculated as = V / I

$$= 0.7 / 10 \text{ mA}$$
$$= 70 \text{ Ohms.}$$

Likewise, if the reverse current and the reverse biasing voltage are given as 0.2 micro-amps and 50 volts, respectively, then, the reverse resistance of the junction is given by

$$50 / 0.2 \text{ micro-amps} = 250 \text{ M-Ohms.}$$

For small signal operations, however, the **incremental resistance**, or **dynamic resistance** is of importance, and is given by a ratio of **change in the voltage** required to produce unit **change in current** around the operating point. It is obvious that this resistance is not constant, since the characteristic is not linear.

From Equation 2.4, it can be shown that the dynamic resistance = $r = \dfrac{dV}{dI}$ can be approximately given by

$$r = \frac{n\,v_T}{I} \tag{2.6}$$

Hence, for $n = 1$, $r = 26 / I$, since
$V_T =$ 26 mV. (I is in mA).

Thus, the resistance, r, will vary from 26 Ohms to 1 Ohm if the current varies from 1 mA to 26 mA.

In most small signal applications, though, it is convenient to use a constant value for this resistance.

PIECEWISE LINEAR REPRESENTATION

Usually for relatively large signal operations, piecewise linear representation produces sufficiently accurate results. Refer to Fig. 2.3. Shown in the figure is a characteristic which indicates zero current upto a cut-in voltage. From this point onwards, the incremental resistance is shown as constant, producing a straight line having a slope equal to $1 / R_f$, and is called a **forward resistance.** (Or resistance of forward biased *p-n* junction).

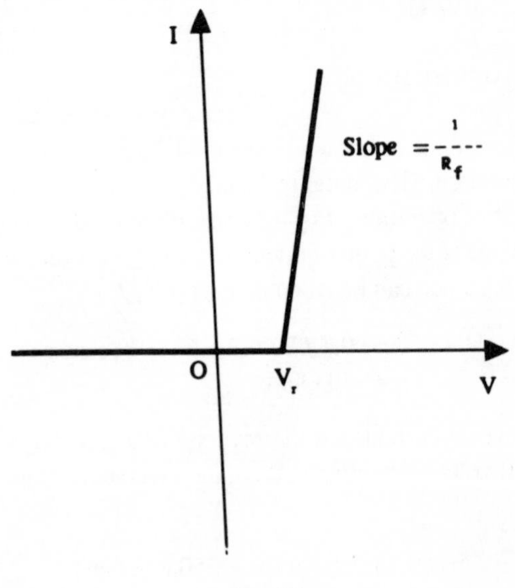

Fig. 2.3

2.2 Zener Diode

Zener diodes have been discussed in Part 1 of this book while discussing the design of zener voltage regulators and the series voltage regulators. A typical circuit and the characteristic is reproduced in Fig. 2.4. Such diodes can be modelled as shown in Fig. 2.5.

As shown in the figure,

where R_z is the zener resistance. This zener resistance produces the slight tilt in the constant current part of the characteristics.

V_z is the zener voltage, for which the diode is rated, and

I_1 is the current source dependent on the current I_1 as shown in the diagram.

This assumes that the load current variations are almost negligible, i.e., any changes in the total current I_1 due to changes in the input voltage variations are reflected in the zener current only, and not in the load current, since the load current will remain almost constant as the output voltage remains essentially constant. The change in the load current due to change in the load resistance can not be accounted for by this model.

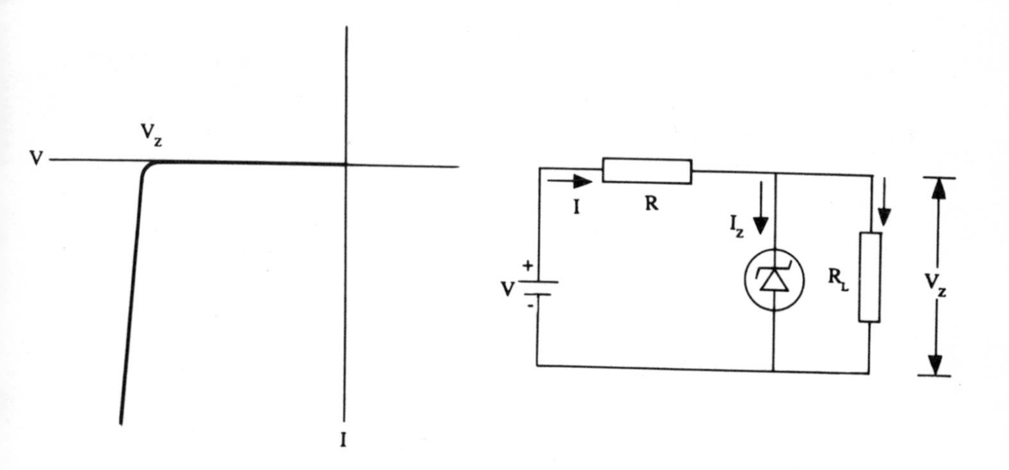

Fig. 2.4

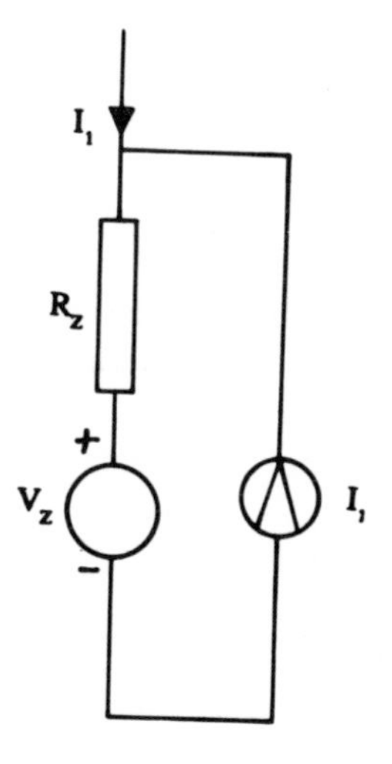

Fig. 2.5

2.3 Bipolar Transistor

A transistor model based on the Ebers - Moll is shown in Fig. 2.6. NPN transistor is assumed for the model. Shown in the model are the following parameters:

1. R_{CC} : This is the collector bulk resistance, and also includes the contact resistance (Typical value is about 10 Ohms.)
2. R_{BB} : The bulk and the base spreading resistance. (Typical value about several tens of Ohms.)
3. R_{EE} : The emitter bulk and contact resistance. (Typical value about 10 Ohms.)

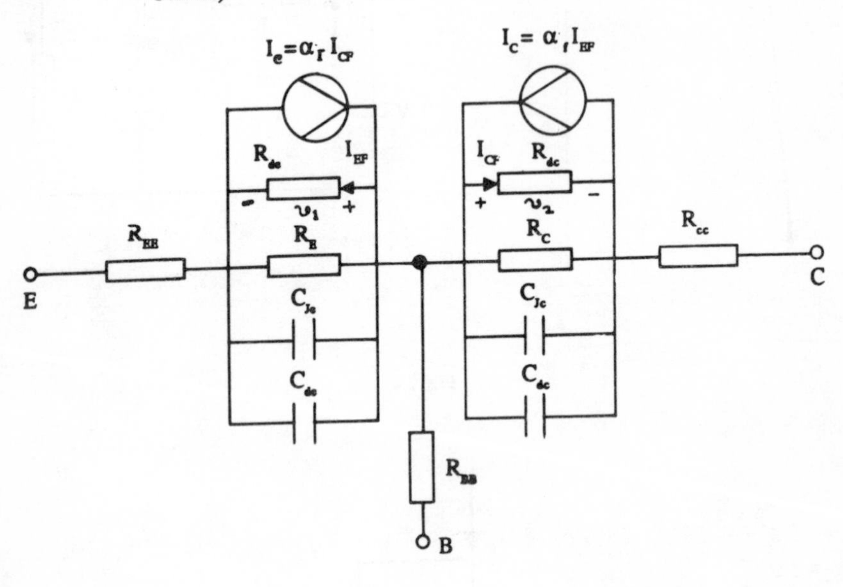

Fig. 2.6

4. R_E & R_C : These are the leakage resistances of emitter-base junction and collector-base-junction, respectively. (Typical values are more than 1 M-Ohm).
5. I_c : The emitter current-controlled current source. This is given by an expression

$$I_c = \alpha_F I_{EF} \tag{2.7}$$

where α_F typically could be about 0.99, and is called normal common-base *dc* current gain. I_{EF} = emitter-base diode forward current.

6. I_e : The collector current-controlled current source. This is given by an expression

$$I_e = \alpha_R I_{CF} \tag{2.8}$$

where α_R typically could be less than 0.5, and is called inverted-mode common-base *dc* current gain.

I_{CF} = collector-base diode forward current.

I_{EF} = emitter-base diode forward current.

7. R_{de} & R_{dc}: These two resistances are the emitter-base and the collector-base diode resistances, as described in Equation 2.1; these diode resistances, are also non-linear. Non-linearity is best described by the following expressions: For R_{de}

$$I_{EF} = I_{ES}\left[\exp\left(\frac{v_1}{M_e V_T} - 1\right)\right] \tag{2.8}$$

For R_{dc}:

$$I_{CF} = I_{CS}\left[\exp\left(\frac{v_2}{M_c V_T} - 1\right)\right] \tag{2.9}$$

where

I_{ES} and I_{CS} are, respectively, emitter-base diode and collector-base diode saturation currents. These are typically much less than a micro-amp, and are about a pico-amp.

M_e and M_c are emission constants for emitter-base diode and collector-base diode, respectively.

The other constants given in the equation have the same meaning as given in Equation 2.1

8. C_{je}: This is an emitter-base junction transition capacitance and is non-linear, given by

$$C_{je} = \frac{D_1}{(V_{te} - v_1)^{ne}} \tag{2.10}$$

where

D_1 = Proportionality constant.

V_{te} = Emitter-base junction contact potential.

n_e = A grading constant (typically between 0.1 and 0.5)

9. C_{jc}: This is a collector-base junction transition capacitance and is non-linear, given by

$$C_{je} = \frac{D_2}{(V_{tc} - v_1)^{nc}} \tag{2.11}$$

where

D_2 = Proportionality constant.

V_{tc} = Collector-base junction contact potential.
n_c = A grading constant (typically between 0.1 and 0.5)

10. C_{de}: The non-linear emitter-base junction diffusion capacitance. The value of this capacitance depends on the charge storage as mentioned in Equation 2.1. Hence, this depends on the value of the emitter-base diode forward current, I_{EF}, which in turn, is given by Equation 2.8. We can write the equation for the capacitance, considering the normal mode gain bandwidth product of the intrinsic transistor; as

$$C_{de} = \frac{I_{ES}}{2\, M_e\, V_T f_n} \exp\left(\frac{v_1}{M_e\, V_T}\right) \tag{2.12}$$

11. C_{dc}: The non-linear collector-base junction diffusion capacitance. The value of this capacitance depends on the charge storage as mentioned in Equation 2.1. Hence, this depends on the value of the collector-base diode junction current, I_{CF}, which in turn, is given by Equation 2.9. We can write the equation for the capacitance, considering the normal mode gain bandwidth product of the intrinsic transistor; as

$$C_{dc} = \frac{I_{CS}}{2\, M_c\, V_T f_n} \exp\left(\frac{v_2}{M_c\, V_T}\right) \tag{2.13}$$

As before, the *dc* model for the bipolar junction transistor can readily be obtained by removing four capacitances from Fig. 2.6. Such a model is given in Fig. 2.7.

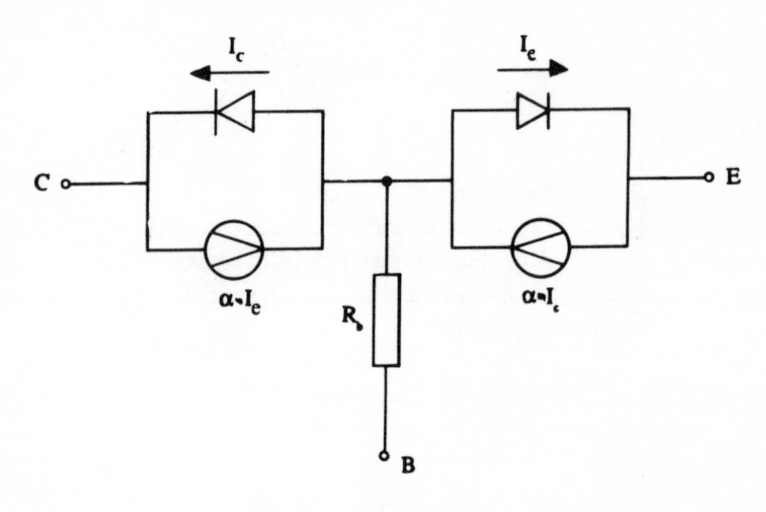

Fig. 2.7

3

Design of Circuits

With the introduction to the models in Chapter 2, and the method of solution of simultaneous equations by nodal analysis in the first Chapter of this part of the book, let us now try to convert some of the popular circuits to networks which can be analysed and designed by using this knowledge. The first circuit to be dealt with is an inverter circuit.

3.1 NOT Gate

The circuit for a simple NOT gate, or the digital inverter circuit is given in Figure 3.1.

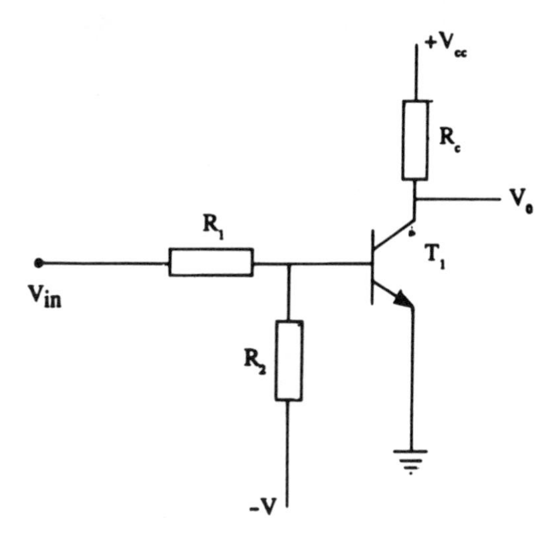

Fig. 3.1

The network for the same, using a typical transistor model, is given in Fig. 3.2. The circuit will be assumed to be working in the linear region, and hence, the presence of all currents shown with appropriate directions will be assumed. Later on, we shall impose conditions such that the circuit can be treated as a digital circuit. Let us write down a set of equations which will fully describe the circuit nodal information.

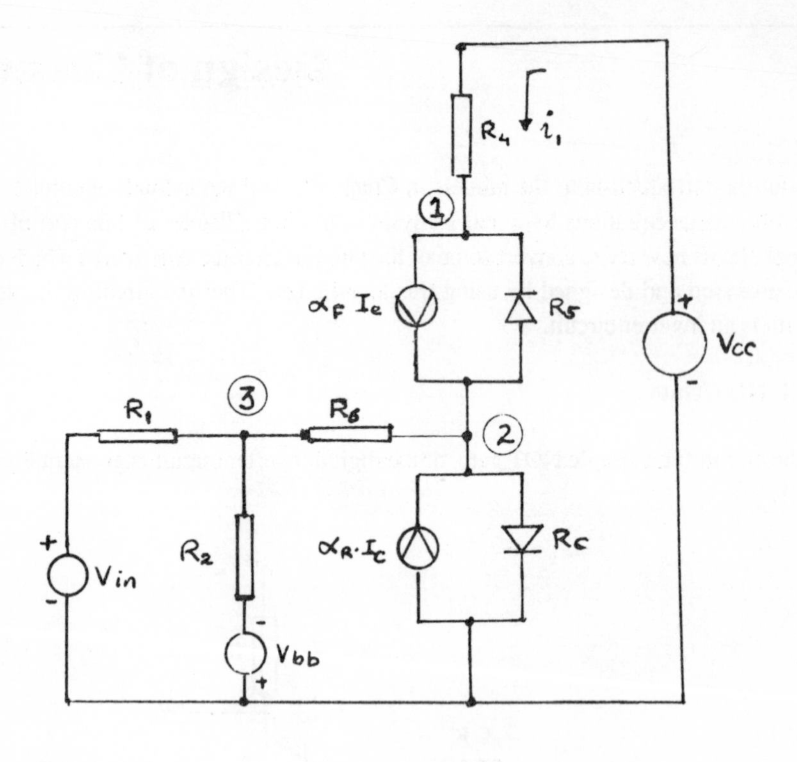

Fig. 3.2

Node No. Equation

1. $$\frac{V_{CC}-V_1}{R_4}+\frac{V_2-V_1}{R_5}-\alpha_F I_e=0$$

2. $$\alpha_F I_e+\frac{V_1-V_2}{R_5}+\frac{V_3-V_2}{R_3}-\frac{V_2}{R_6}+\alpha_r I_c=0$$

3. $$\frac{V_{in}-V_3}{R_1}-\frac{V_3+V_{BB}}{R_2}+\frac{V_2-V_3}{R_1}=0$$

This set of equations can be easily solved by the computer program given in the first Chapter. It should be clear, however, that since we have written three equations, the number of unknown variables should not exceed three. In case of more than three unknown variables, obviously we shall have to define more equations equal to the number of unknown variables.

In case, the circuit is to be used as a digital circuit, it will be better to tie up resistance R_1 to supply voltage, V_{CC}. Also, the voltage V_2 will be equal to the saturation value specified for the given transistor. The second calculation will involve an equation set with voltage V_{in} set equal to zero, forcing voltage V_2 to be equal to V_{cc}.

Let us take one more example where we shall take a similar circuit, but where the transistor is not forced into saturation. This will permit us to modify the transistor model.

3.2 Transistor Common-Emitter Voltage Amplifier

The circuit given in Fig. 3.3 represents a common-emitter amplifier. The circuit uses a single transistor. When the transistor is replaced by the model discussed earlier, the corresponding circuit appears as given in Fig. 3.4. Notice the modified model of the transistor. This model represents the transistor in the active region and is fairly precise. The circuit has six nodes, including the zero node, and there are five unknowns, namely the voltages at nodes 2, 3, 4 and 5, and the current i_1 from the voltage supply V_1.

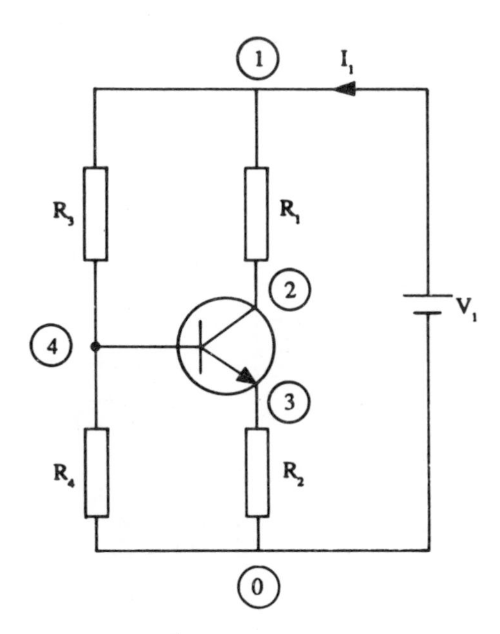

Fig. 3.3

Assuming that the voltage at node 0 is 0 V, the nodal equations can be written as under.

Node No.	Equation
1.	$0 = i_1 + \dfrac{V_4 - V_1}{R_3} + \dfrac{V_2 - V_1}{R_1}$
2.	$0 = \dfrac{V_1 - V_2}{R_1} + \dfrac{V_5 - V_2}{R_r} - I_e$
3.	$0 = \dfrac{0 - V_3}{R_2} + I_e$
4.	$0 = \dfrac{V_1 - V_4}{R_3} + \dfrac{0 - V_4}{R_4} + \dfrac{V_5 - V_4}{R_b}$
5.	$0 = \dfrac{V_4 - V_5}{R_b} + \dfrac{V_2 - V_5}{R_r} - I_e + \alpha I_e$

where v_n is the voltage at node n.

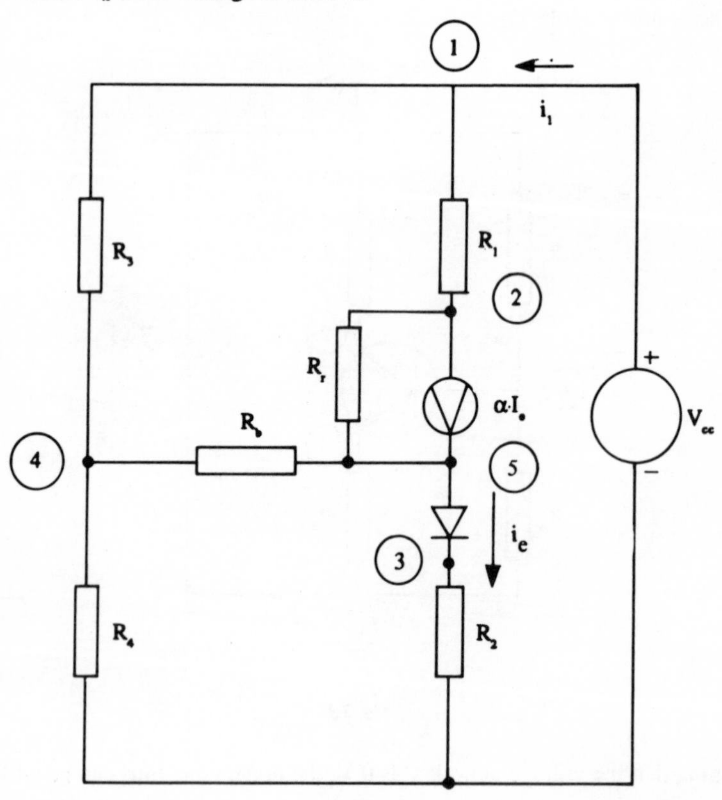

Fig. 3.4

Totally, there are five equations and five unknowns, of which two are linear and rest are non-linear. The above equations, written in the form of matrices, can be solved as mentioned in the first Chapter.

3.3 Emitter - Follower Type of Voltage Regulator

Let us now take a slightly more complex circuit than the ones we have considered so far. Given in Fig. 3.5 is a zener voltage regulator using an emitter-follower.

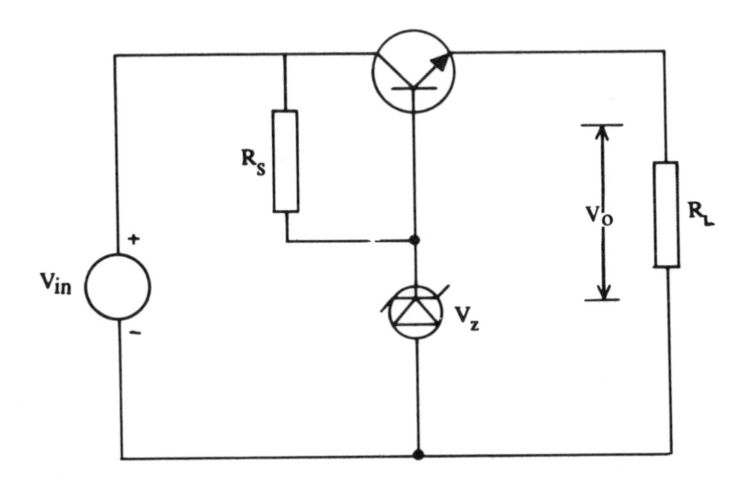

Fig. 3.5

The operation of the circuit is described in Chapter 4 of Part 1 of the book. Shown in Fig. 3.6 is a network using a classic model of the transistor and a modified model of the zener diode. The transistor model can be further modified between nodes 2 and 3, as was done in Fig 3.4. With these modifications, it becomes a simple network with four nodes (plus the reference node) for which a set of equations can be written and analytical solution obtained, following the same method as was used in previous examples. It should be noted that the zener model is valid only for the condition after breakdown. This again is an approximate model, and does not take into account the change in the zener current due to change in the output voltage. In general, we can say that the model is good enough only where current variation in the diode current is negligible. A dependent current source as shown by the dashed line in the diagram must be used, increasing the network complexity.

We shall now discuss the design methodology of some of the circuits we have discussed with the help of the computer. In these designs we shall demonstrate the use of the computer to reduce the drudgery of remembering and solving common equations and also the steps to be followed in a practical design. First, we shall see the design considerations, and then write a program for the same.

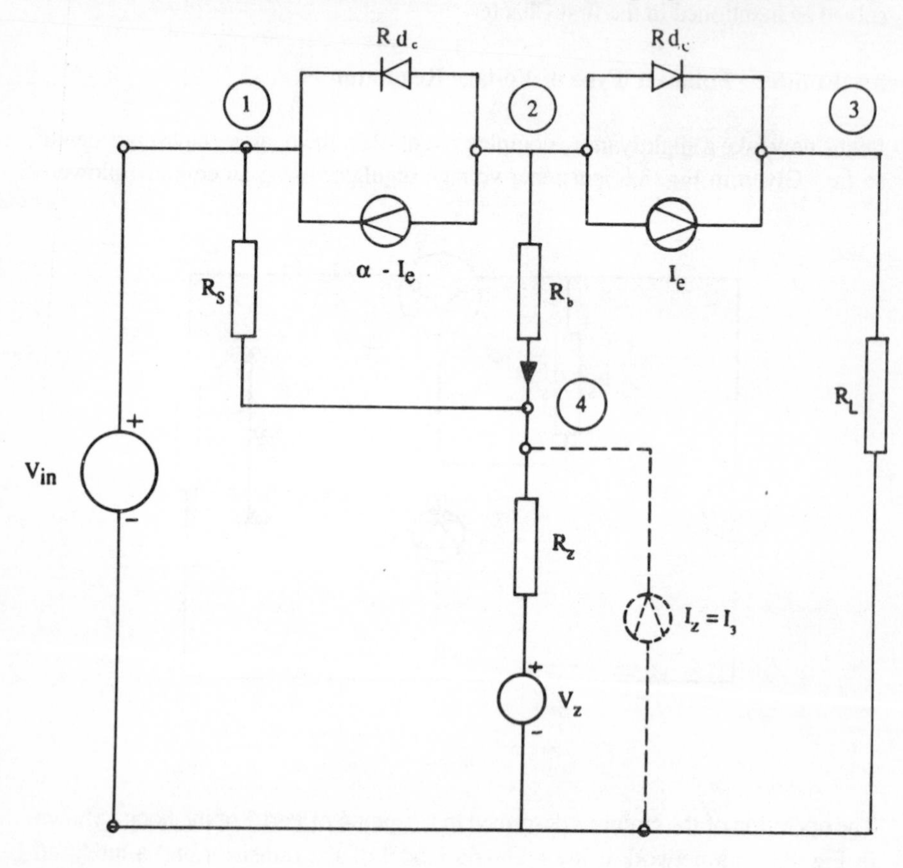

Fig. 3.6

3.4 Design of Inverter Circuit (NOT)

In the circuit in Fig. 3.1, the transistor is essentially used as a switch. When the transistor is used as a switch, it will have to have only one of the two operating points, either at cut off or at saturation, i.e., at a point A or at a point B as in Fig. 3.7. Hence, our strategy should be that, given the supply voltage, first the resistance, R_c, should be selected.

Before we can select R_c, we will have to consider and select the value of current I_c, which should flow through the transistor during saturation condition. For this purpose, in absence of other specifications, $I_{c(sat)}$ should be selected as between about 5% and 10% of $I_{c(max)}$ specified by the transistor manufacturer. For this

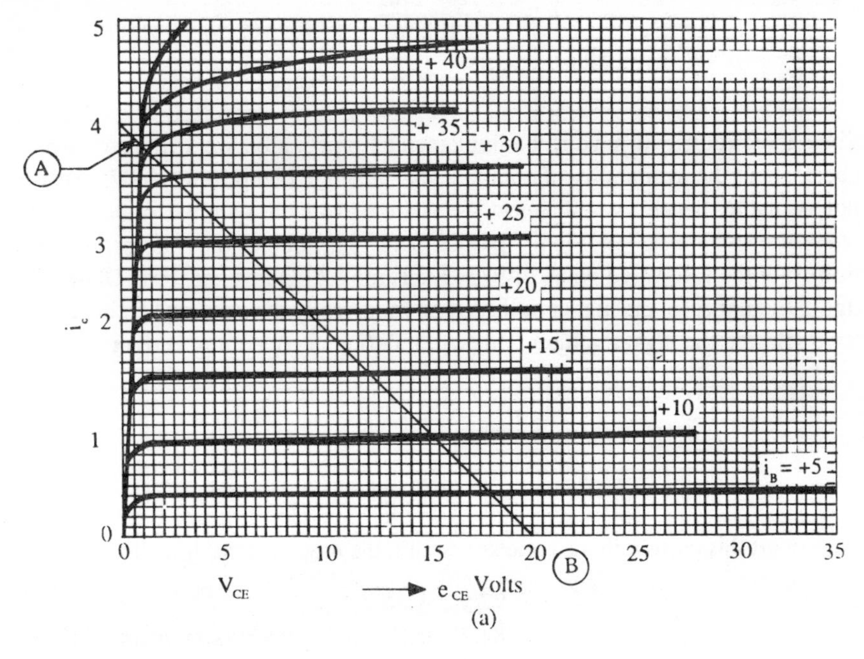

Fig. 3.7

purpose, we shall have to first begin our search for the transistor for possible use in the circuit.

Since no limitations are placed on the type to be selected, we can go ahead and select one of the many switching transistors available, or for that matter, any small signal general purpose transistor. (If rise time, fall time and frequency of operations are specified, we will have to select a specific transistor for the purpose.) Let us select BC147A. This transistor has the following relevant specifications:

$$\begin{aligned} I_C(\text{max}) &= 200 \text{ mA at } 25°\text{C.} \\ h_{FE}(\text{min}) &= 115 \\ h_{FE}(\text{max}) &= 220 \\ V_{CE}(\text{sat}) &= 0.25. \end{aligned}$$

Since $I_C(\text{max}) = 100$ mA, we shall select the operating current as about 10% of $I_C(\text{max}) = 10$ mA.

Hence, if supply voltage V_{CC} is 10 Volts,

$$R_c = \frac{V_{CC} - V_{CE(\text{sat})}}{I_{C(\text{sat})}} \qquad \text{(i)}$$

$$= \frac{10 - 0.25}{10 \text{ mA}}$$

$$= 0.975 \text{ K} = 975 \text{ ohms}$$

Since the circuit will have to be implemented practically, we will also have to check about the availability of components. A single component of 975 ohms is not available, hence we will select the nearest available nominal value for this resistance, which is 1 k-Ohm. (Notice here that, among the values available nearest to the calculated value of 975 Ohms, are 820 Ohms and 1000 Ohms. In this particular case, the higher or the lower value really does not matter, hence we have selected the higher value.) This will redefine the operating current as

$$I_{c(sat)} = \frac{10 - 0.25}{1.0 \text{ K}}$$

$$= 9.75 \text{ mA.}$$

To support this current in the collector circuit, the base-current, I_b, will have to be

$= 9.75 \text{ mA} / h_{FE(min)}$
$= 84.8$ micro-Amp. at the maximum and
$= 9.75 \text{ mA} / h_{FE(max)}$
$= 9.75 \text{ mA} / 220$
$= 44.3$ micro-amps at the minimum.

To be on the safer side, the design should employ the minimum value of h_{FE} and not the maximum or typical, for, otherwise, any change in the current gain due to temperature change or due to device replacement would require fresh design. This base current is derived from the input voltage when it is held at **1** level. Normally, **1** level would mean voltage at the supply level. In our case we shall assume it to be 10 V. Current I_1 will flow through the resistance R_1. At base of the transistor, this current splits up in two parts, into current I_2, which passes through resistance R_2 to the biasing supply, and the remaining current, $(I_1 - I_2)$, which becomes the base-current I_b.

One more point which should be kept in mind is that the resistance selected will have a certain tolerence, that is to say that if we select a 1 K-Ohm resistance, the actual value may differ from this nominal value by about ±10 % (for economically priced commercially available resistances), i.e., the actual value of the resistance of 1 K value could be anywhere between 1.1 K-Ohms (10% more) and 0.9 K-Ohm (10% less). Let us see the effect of this variations from among the resistances that we are going to select.

R_c : In case, this resistance is **less** by 10%, current I_c will increase, thereby requiring **increased** I_b, base-current requirements.

R_1 : If this resistance is **more** than the nominal value, current I_1 will reduce, **reducing** the base-current actually flowing into the base-emitter junction.

R_2 : If this resistance is **less** than the nominal value, current I_2 will increase, **reducing** base-current actually flowing into the base.

Assuming the worst case, where all the above effects are cumulative, the base-current flowing into the base will not be sufficient to keep the transistor in saturation. The supply voltage variations, as also the biasing voltage variations, will also effect the base current requirements. However, assuming that fairly regulated power supplies are available, the base-current requirements will still have to be enhanced by **at least 30%**, or more, depending on the number of resistances involved.

The ratio $\frac{I_{b(actual)}}{I_{b(required)}}$ is called the **Over Drive Factor (O.D.F)**.

Keeping this in mind, the base current requirement of 84.8 micro-amps will have to be increased to about

$$1.3 \times 84.8 = 0.11 \text{ mA.}$$

$$= I_1 - I_2 \qquad \text{(a)}$$

Assuming that the base-emitter junction voltage, when transistor is in saturation, is about 0.7 volts, (or whatever is specified by the transistor manufacturer), we have,

$$I_1 = \frac{V_{in} - V_{BE(sat)}}{R_1} \qquad \text{(ii)}$$

$$= (10 - 0.7)/R_1$$

$$= \frac{9.3}{R_1} \qquad \text{(b)}$$

Likewise

$$I_2 = \frac{V_{BE(sat)} - V_{BB}}{R_2} \qquad \text{(iii)}$$

$$= [0.7 - (-5)]/R_2$$

$$= \frac{5.7}{R_2} \qquad \text{(c)}$$

Hence, $$I_B = I_1 - I_2$$

$$= \frac{9.3}{R_1} - \frac{5.7}{R_2} \tag{d}$$

$$= 0.11 \text{ mA from (a).}$$

Equation (d) contains two unknown variables, R_1 and R_2. Another equation, therefore, is required in order to facilitate finding and determining the values for these resistances. When the input is at zero level, 0 Volt in our case, it should keep the transistor in cut-off state. The negative bias used should help in keeping the base-emitter junction reverse biased, to keep transistor effectively off. Hence, when V_{in} is zero, the base voltage will be equal to

$$V_B = \frac{R_2}{R_1 + R_2} \times V_{BB} \tag{iv}$$

Since the emitter is at a ground level, this voltage will be effectively, V_{BE}, which will be negative, since the bias voltage V_{BB} is negative.
In our case,

$$V_{BE} = \frac{R_1}{R_1 + R_2} \times (-5) \tag{e}$$

This voltage needs to be only slightly negative, say about one or two volts, to keep the transistor off. We shall assume this voltage to be about (–1.5 V), giving us a relation between R_1 and R_2, as

$$-1.5 = \frac{R_1}{R_1 + R_2} \times (-5)$$

i.e., $$\frac{R_2}{R_1} = 2.33 \tag{f}$$

From Equation d, we have

$$0.11 = \frac{9.3}{R_1} - \frac{5.7}{2.33 R_1}$$

$$= \frac{6.85}{R_1}$$

From which **R_1 = 62.27 k-Ohms, say 62 K-Ohms.**

and **R_2 = 144.46 k = say 150 k-Ohms.**

Notice the selection of the nominal values of the resistances, which is done to effectively increase the base-current.

COMPUTER PROGRAM FOR THIS DESIGN

The above design is now implemented using a Pascal program.
Refer to Program P3.

Some of the points need explanation.

In the variable and constant declarations, O.D.F, the over-drive factor, has been assigned a value 1.3, as explained in the previous design. $V_{BE(sat)}$ is assumed as 0.7 and base-voltage for cut-off is assumed as (−1.5), and are declared to have these values.

VAR declaration section declares two files. The first is declared to have a set of transistor numbers. Each of the transistors is listed along with other pertinent data, like maximum current, minimum h_{FE}, etc.

The second file is the resistor file, which is supposed to contain a list of the nominal values of the resistances available from which the selection will be made by the program. This selection is done by the program through a procedure written for the purpose, called **procedure resistor**.

The values of resistances available are assumed to be **1, 1.5, 2.2, 2.7, 3.3, 3.9, 4.7, 5.1, 5.6, 6.2, 6.8, 7.5, 8.2 and 10 Ohms**, and the above values multiplied by powers of ten. The calculated values are, therefore, reduced to a format $\boldsymbol{R \times 10^d}$, where R is less than 10 Ohms. A 975 Ohms resistance is reduced by the program to 9.75×10^2, hence $R = 9.75$ and $d = 2$.

The program then proceeds to read two values from the file, and compares them with the calculated values. It enters into a loop till it finds two values ***rn*** and ***rp*** such that ***rn* is greater** than, and ***rp* is less** than the calculated values. For the above example, *rn* will be equal to 10, and *rp* will be equal to 8.2. These two values are required, since in some cases we would like to select a value which is greater than the calculated value (like the value of resistance R_2, in our case) and in some other cases we would like a value which is less than the calculated value (like for resistance R_1, in our case).

The main block of the program then proceeds to ask the user about the values of the supply voltage V_{CC}, bias voltage V_{BB}, and the transistor to be used. (Note that the program can be easily modified to let it search for the transistor number, were it fed with the operating current.) The program then reads the relevant information regarding the transistor and prints on the screen. It, then, calculates the value of the resistance R_c, taking operating current as about 10% of the $I_{c(max)}$, and promptly calls for the procedure which will select the nominal value **which is higher than the calculated value** by selecting *rn*, and not *rp* (which is lower than the calculated value.) The program then, calculates values for resistances R_1 and R_2 and, through the selection procedure, finds from the file a value which is greater than the calculated value (for R_2), and a value which is less than the calculated value (for R_1).

Program-P3

```
PROGRAM logic_design (input,output)
(* THIS PROGRAM IS USED TO DESIGN LOGIC NOT GATE *)

CONST
    odf   =  1.3;
    Vbes  =  0.7;
    Vb1   =  -1.5;
 VAR
    icm,x,Vces,hfm,rp,rn,ics,rc,iba,n,r1,r2     :  real;
    Vbb,Vcc,e,i                                 :  integer;
    transfile,resfile                           :  text;
    tno,tno1,icm1,hfm1,Vces1                    :  string[6];
    tranline                                    :  string[30];

    PROCEDURE resistor (ro : real; VAR rp,rn : real);
     (* to find standard resistor value *)
    VAR
        d         :  integer;
        r         :  real;
        resline   :  string[7];
        rp1,rn1   :  string[3];
    BEGIN
    d := 0;
    WHILE ro >= 10 DO
    BEGIN
        ro  :=  ro/10;
        d   :=  d + 1;
    END;

    reset(resfile);
    readln(resfile,resline);
    rpl := copy(resline,5,2);
    val (rpl,rp,e);
    readln(resfile,resline);
    rnl := copy(resline,5,2);
    val(rn1,rn,e);

    WHILE rn < ro DO
    BEGIN
        rp := rn;
        readln(resfile,resline);
```

```
            rn1 := copy(resline,5,2);
            val(rn1,rn,e);
        END;

        rp := rp * exp(d * ln(10));
        rn := rn * exp(d * ln(10));
        END;        (* end of resistor *)


BEGIN       (* main action block *)
        ASSIGN (transfile,'trans.pas');
        ASSIGN (resfile,'res.pas');
        writeln('                  DESIGN OF LOGIC NOT GATE ');
        writeln('                  ------------------------ ');
        writeln;
        writeln;
        writeln(' THIS PROGRAM IS USED TO DESIGN A NOT ');
        writeln(' GATE USING BJT & RESISTORS. ');
        writeln;
        write('INPUT VALUE OF Vcc & Vbb  :');
        readln(Vcc,Vbb);
        writeln;
        write(' INPUT TRANSISTOR NUMBER:');
        readln(tno);
        rp := 0;
        rn := 0;
        BEGIN          (* accessing the transistor file *)
            reset(transfile);
            WHILE (tno <> tno1) DO
            BEGIN
                readln(transfile,tranline);
                tno1 := copy(tranline,5,6);
            END;

            writeln;
            write('TRANSISTOR SELECTED FROM DATASHEET IS :  ');
            writeln(tno);
            writeln;
            icm1 := copy(tranline,12,3);
            val(icm1,icm,e);
            writeln('Ic(max) OF THE TRANSISTOR IS    :',icm:4:2,'mA');
            writeln;
            Vces1 := copy(tranline,16,4);
            val(Vces1,Vces,e);
```

```
        writeln('Vce(sat) OF THE TRANSISTOR IS   :',Vces:4:3,'V');
        writeln;
        hfm1 := copy(tranline,21,3);
        val(hfm1,hfm,e);
        writeln('THE VALUE OF hFE(min) IS   :   ',hfm:3:1);
        writeln;
    END;
        readln;
  BEGIN
    (* this designs a NOT gate *)

    ics   :=  0.1 * icm;          { calculating Rc }
    rc    :=  (Vcc – Vces)/ics;
    writeln('              *   NOT GATE DESIGN *');
    writeln('              ------------------------        ');
    writeln;
    writeln('CALCULATED VALUE OF Rc IS   :',rc:6:2, 'kohms');
    resistor(rc,rp,rn);
    writeln('STANDARD VALUE OF Rc IS  :    ',rn:6:2, 'kohms');
    rc    :=  rn;
    ics   :=  (Vcc – Vces)/rn;
    writeln;
    writeln('RECALCULATED VALUE OF Ic(sat) IS   :',ics:3:2,' mA')


    iba   :=      odf * (ics /hfm);  { calculating R1 & R2 }
    x     :=      (Vb1 + Vbb)/1.8;
    r1    :=      (1.3 – (Vbes + Vbb)/x)/iba;
    r2    :=      abs(x * r1);
    writeln('CALCULATED VALUE OF R1 IS     :',r1:6:2,' kohms');
    resistor(r1,rp,rn);
    writeln('STANDARD VALUE OF R1 IS  :    ',rp:6:2,' kohms');
    writeln;
    writeln('CALCULATED VALUE OF R2 IS     :',r2:6:2,' kohms. ');
    resistor(r2,rp,rn);
    writeln('STANDARD VALUE OF R2 IS  :    ',rn:6:2,' kohms. ');
    writeln;
    writeln(' DESIGN COMPLETE. ');
    writeln(' --------------- ');
  END;
  readln;
END.
```

The design is thus complete.

3.5 Design of Bistable Multivibrator

Figure 3.8 shows a bistable multivibrator. This circuit is essentially made up of two inverter circuits designed above, connected back-to-back. V_{in} of the inverter is now derived from the collector of the other transistor. This means that if transistor T_1 is on and in saturation, its output voltage at the collector being almost zero (about 0.3 V), and will not be able to supply the base current for the other transistor, T_2. The transistor T_2 will, therefore, remains off. The voltage at collector of transistor T_2 will be approximately equal to V_{CC}, since the only current that flows from the resistance R_c of this transistor, is current I_1 going to supply the base-current of the transistor T_1. This high voltage will be the input voltage, V_{in}, for transistor T_1, which should keep transistor T_1 in saturation.

The design procedure to be adopted for this circuit should be similar to what we have seen Article 3.4 above. Let us summarise the methodology.

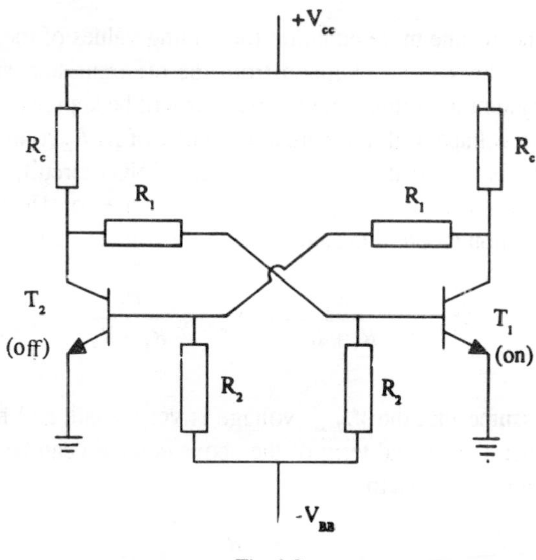

Fig. 3.8

Step 1. Given the supply voltage V_{CC}, and the biasing voltage V_{BB}, we can proceed to select the transistor to be employed from the required conditions.

Step 2. We select the operating current for the transistor from the conditions given. In absence of such conditions, we select $I_{c(sat)}$ as about 5% to 10% of the $I_{c(max)}$ of the selected transistor.

Step 3. We calculate resistance R_c, as was done in design of NOT gate, by

$$R_c = \frac{V_{CC} - V_{CE(sat)}}{I_{c(sat)}}$$

and find the standard available value for the same. We recalculate the saturation current $I_{c(sat)}$. This value of the actual current will have to be used for further calculations.

Step 4. We determine minimum value of the base-current needed to support $I_{c(sat)}$, with the use of $h_{FE(min)}$. Assuming an O.D.F, we select a higher value for I_b.

Step 5. This current I_b is given by $(I_1 - I_2)$, where

$$I_1 = \frac{V_{CC} - V_{BE(sat)}}{R_1}$$

and

$$I_2 = \frac{V_{BE(sat)} - (-V_{BB})}{R_2}$$

Step 6. Step 5 needs one more equation for finding values of the resistances R_1 and R_2. This can be obtained from the off transistor, since the base-emitter junction of this second transistor will be kept reverse biased by a negative voltage with the potential divider of R_1 R_2 combination. This, however, is slightly different than in case of NOT circuit, since zero level voltage in this case is equal to $V_{CE(sat)}$ and not zero. Hence, we employ superposition theorem in finding the base voltage. Thus;

$$V_{BE(off)} = \frac{R_2}{R_1 + R_2} \times (-V_{BB}) + \frac{R_1}{R_1 + R_2} \times V_{CE(sat)}$$

If we assume that the $V_{CE(sat)}$ voltage is very small, and hence is negligible, then the second term of the above equation can be discarded and the equation reduces to

$$V_{BE(off)} = \frac{R_2}{R_1 + R_2} \times (-V_{BB})$$

Or

$$R_1 = R_2 \times \frac{-V_{BB} - V_{BE(off)}}{V_{BE(off)}}$$

$$= n \times R_2$$

Step 7. From the above two steps we can now calculate values of resistances R_1 and R_2.

It is imperative that, since the values of the components obtained in the above design are converted to values which are commercially available, the designed circuit should be rechecked for the performance. The computer program, therefore, should include this facility.

COMPUTER PROGRAM FOR DESIGN OF BISTABLE MULTIVIBRATOR

The computer program in PASCAL for designing of the bistable circuit is given in Program P4. This program proceeds in exactly the same way as that written for the NOT circuit. Explanation given for NOT circuit Program P3, therefore, holds good for this circuit as well. There are a couple of points, however, which need elaboration.

1. The O.D.F declared is 1.5 instead of 1.3, as in previous case. Strictly speaking, this is not necessary. However, in this circuit there is one more resistance which may affect the value of current I_1. Besides, there is nothing wrong in being a little more cautious.

2. The second block of the program from end marked ***verification of the circuit*** verifies whether values of the components chosen will give us a bistable circuit. What it does is to calculate the voltage v_{b2}, which is a voltage at the base of the off transistor. This voltage which was assumed as -1.5 V, may be actually different than the calculated value, though the actual value of the same is not critical. If it is negative enough to keep the transistor off, we ignore the actual value. The other value of the voltage checked is at the collector of the ON transistor. This is just a check to see whether the transistor has gone into saturation. The condition checked is not sufficient to prove conclusively the saturation condition of the transistor, but, in almost all cases, is good enough.

3. The last block of the program calculates the value of the load resistance that our designed circuit will support. The off transistor, having almost entire V_{CC} available at the collector, should be able to supply the load current to external circuit.

Program - P4

```
PROGRAM bistable_multi(input,tranfile,resfile,output);
(* THIS PROGRAM DESIGNS A BISTABLE MULTIVIBRATOR *)
const
      odf   =  1.5;
      Vbes  =  0.7;
      Vb1   =  -1.5;
VAR
      icm,Vces,hfm,rp,rn,ics,rc,ibm,iba,n,r1,r2   :  real;
      Vb2,V1,Vc2,i2,i1,Vc1,iL,rL                  :  real;
      Vop,Vbb,Vcc,i,e                             :  integer;
      tranfile,resfile                            :  text;
      tno,tno1,icm1,hfm1,Vces1                    :  string[6];
      tranline                                    :  string[30];

PROCEDURE resistor (ro : real; var rp,rn : real);
(* TO FIND THE STANDARD RESISTOR VALUE *)

VAR  d        :  integer;
     r        :  real;
     rp1,rn1  :  string[3];
     resline  :  string[7];

 BEGIN
  d := 0;
  WHILE ro >= 10 DO
   BEGIN
    ro   :=  ro/10;
    d    :=  d + 1;
   END;
  reset(resfile);
  readln(resfile,resline);
  rpl := copy(resline,5,3);
  val(rpl,rp,e);
  readln(resfile,resline);
  rnl := copy(resline,5,3);
  val(rn1,rn,e);
  WHILE rn < ro DO
   BEGIN
    rp := rn;
    readln(resfile,resline);
    rn1 := copy(resline,5,3);
```

```
    val(rn1,rn,e);
   END;
  rp := rp * exp(d * ln(10));
  rn := rn * exp(d * ln(10));
 END; (* end of resistor *)


BEGIN (* main *)
 ASSIGN(tranfile,'trans.pas');
 ASSIGN(resfile,'res.pas');
 writeln('BISTABLE MULTIVIBRATOR DESIGN ');
 writeln('---------------------------- ');
 writeln;
 write('INPUT VALUES OF Vcc AND Vbb :  ');
 readln(Vcc,Vbb);
 writeln;
 write('INPUT TRANSISTOR NUMBER:');
 readln(tno);
 rp := 0;
 rn := 0;


  BEGIN              (* accessing transistor file *)
   reset(tranfile);
   WHILE (tno <> tno1) DO
    BEGIN
     readln(tranfile,tranline);
     tno1 := copy(tranline,5,6);
    END;
   writeln;
   writeln('TRANSISTOR SELECTED FROM DATASHEET IS:'tno);
   writeln;
   icm1 := copy(tranline,12,3);
   val(icm1,icm,e);
   writeln('Ic(max) OF THE TRANSISTOR IS       :    ',icm:4:2,'mA');
   writeln;
   Vces1 := copy(tranline,16,4);
   val(Vces1,Vces,e);
   writeln('Vce(sat) OF THE TRANSISTOR IS      :    ',Vces:4:3,'V');
   writeln;
   hfm1 := copy(tranline,21,3);
   val(hfm1,hfm,e);
   writeln('THE VALUE OF hFE(min) IS           :    ',hfm:3:1);
```

```
 writeln;
END;


 BEGIN (* calculating value of Rc *)
  ics := 0.1 * icm;
  rc  := (Vcc – Vces)/ics;
  writeln('CALCULATED VALUE OF Rc IS    :', rc:6:2,' ohms');
  resistor(rc,rp,rn);
  writeln;
  writeln('STANDARD VALUE OF Rc IS  :    ',rn:6:2,' kohms');
  rc  := rn;
  ics := (Vcc – Vces)/rn;
  writeln;
  writeln('RECALCULATED VALUE OF Ic(sat) IS:',ics:3:2,' mA');
 END;

 BEGIN(* calculating value of R1 and R2 *)
  ibm := ics/hfm;
  iba := odf * ibm;
  n   := (Vbb – Vn1)/Vb1;
  r1  := (n * (Vcc – Vbes) – Vbb – Vbes)/(n * iba);
  writeln;
  writeln('CALCULATED VALUE OF R1 IS    :', r1:6:2,' kohms');
  writeln;
  resistor(r1,rp,rn);
  writeln('STANDARD VALUE OF R1 IS  :    ',rp:6:2,' kohms');
  writeln;
  r2  := n * r1;
  r1  := rp;
  writeln('CALCULATED VALUE OF R2 IS    :',r2:6:2,' kohms');
  writeln;
  resistor(r2,rp,rn);
  writeln('STANDARD VALUE OF R2 IS  :    ',rn:6:2,' kohms');
  r2  := rn;
 END;
 readln;


 BEGIN(* verification of the circuit *)
  Vb2 := (Vbb * r1)/(r1 + r2);
  writeln('VERIFICATION OF THE CIRCUIT ');
  writeln('--------------------------- ');
```

```
  writeln;
  writeln('BASE VOLTAGE OF OFF TRANSISTOR IS:',Vb2:4:2, 'Volts');
  Vc2 := (Vcc – ics*rc);
  writeln;
  writeln('COLLECTOR VOLTAGE OF ON TRANSISTOR IS:',Vc2:4:2,
'Volts');
 END;


 BEGIN (* to find load resistance *)
  writeln;
  writeln('CALCULATING LOAD RESISTANCE');
  writeln('---------------------------');
  writeln;
  i2   := (Vbb + Vbes)/r2;
  i1   := ibm + i2;
  Vc1  := i1 * r1 + Vbes;
  iL   := (Vcc – Vc1 – i1*rc)/rc;
  rL   := Vc1/iL;
  writeln('CALCULATED VALUE OF RL IS:',rL:6:2,' kohms');
  writeln;
  resistor(rL,rp,rn);
  writeln('STANDARD VALUE OF RL IS:',rn:6:2,' kohms');
 END;

 writeln;
 writeln('DESIGN COMPLETE.');
 writeln('----------------');
 close(tranfile);
 close(resfile);
 readln;
END.
```

Any current supplied to the load resistance R_L will have to pass through the collector resistance R_c, creating a voltage drop across it. The reduced voltage available at the collector of this transistor will reduce the supply of current I_1, which in turn, will reduce the base current for the ON transistor, since current I_2 will remain essentially constant. (Refer to the equation given in Step 4.) Hence, there is a limit beyond which the current to the load resistance R_L can not be increased, since this, at some point, will force ON transistor to enter into the active region of operation. The maximum current can be calculated in the following way.

$$I_{1(\text{minimum})} \text{ required} = I_{B(\text{min})} + I_2$$

This current creates a voltage drop in the resistance R_1 equal to $I_{1(\text{min})} \times R_1$. Refer to Fig. 3.9.

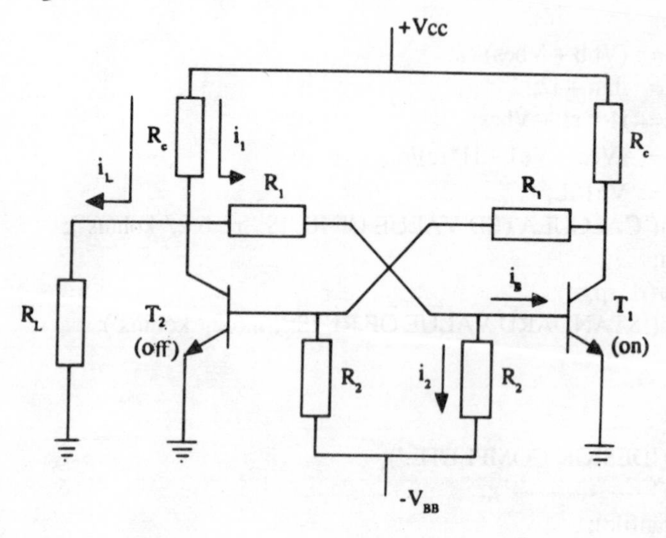

Fig. 3.9

Thus, voltage at the collector of the OFF transistor should be = $V_{BE(\text{sat})} + (I_{1(\text{min})} \times R_1) = V_{c(\text{off})}$

Hence the load current

$$I_L = \frac{V_{CC} - V_{c(\text{off})} - I_1 \times R_c}{R_c}$$

3.6 Design of Monostable Multivibrator

In bistable circuit, both digital inverters are connected with each other by direct coupling. The collector of each of the transistors is connected to the base of the other transistor through resistance R_1. This resistance essentially supplies the base-current for the transistor which is ON, and also forms part of the network with R_2 to provide a fixed negative voltage to keep base-emitter junction of the other transistor to keep it OFF. In a monostable circuit, one of the *dc* coupling is replaced by a capacitor coupling. Refer to Fig. 3.10. Capacitor C, replaces resistance R_1, connected between collector of transistor T_2 to the base of transistor T_1. To supply base-current to transistor T_1 when it is ON, a resistance, R is connected between its base and the supply voltage V_{CC}.

The action of the circuit, briefly, is as follows:

When the supply is switched on, both the transistors should receive base-current. Transistor T_2 gets its base-current through the resistance R_1, as mentioned in the design of the inverter circuit. Transistor T_1 gets its current from the resistance R. However, initially capacitor C is in relaxed condition. Hence, when the circuit is switched ON, the condensor presents a short. Due to this, a large current is supplied to the base of transistor T_1, *albeit* for a very small time, through resistance R_c. The magnitude of this current is considerably larger than a steady-state current which is supplied by resistance R. (Resistance R could be as large as about 100 times the resistance R_c.) Due to this large current, transistor T_1 is forced into saturation first, which in turn will deny current to the base of transistor T_2. This is a stable state. During this state, capacitance C gets charged to almost supply voltage V_{CC}, getting its charge through resistance R_c (of T_2) and the forward biased the base-emitter junction of transistor T_1.

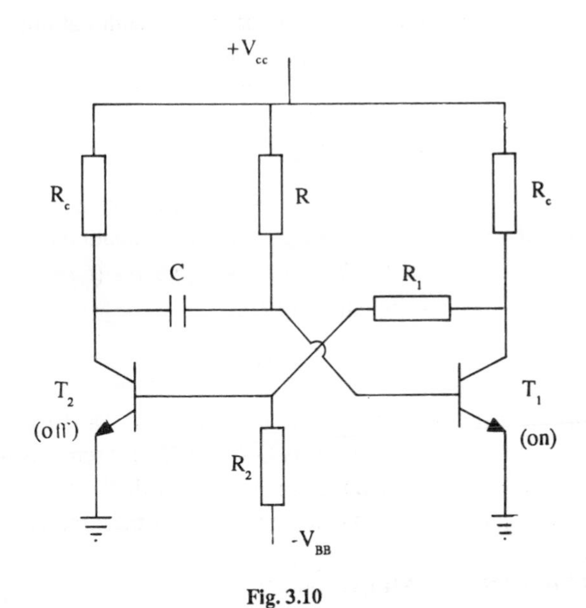

Fig. 3.10

When an external excitation is provided in the form of positive pulse to transistor T_2 to turn it ON (or a negative pulse to transistor T_1 to turn it off), voltage at the collector of transistor T_2 falls. This sudden fall in the voltage Vc_2 is almost entirely passed on to the base of transistor T_1, since the capacitor can not change voltage across itself instantly. This forces transistor T_1 to turn-off, forcing its collector voltage V_{c1} to increase almost to V_{CC}. This increase in the voltage permits the transistor T_2 to turn ON.

The capacitor voltage, which was $+V_{CC}$ at the collector-end and almost zero ($V_{BE1(sat)}$) at the base-end, is now applied; a voltage which is almost zero ($V_{CE1(sat)}$) at collector-end and the base-end now gets connected to a positive V_{CC}. This, therefore, starts charging the capacitance in the opposite direction, forcing base-end voltage to increase exponentially with a time constant ($R \times C$). When the base-voltage of transistor T_1 becomes more than the cut-in voltage, it permits transistor T_1 to turn ON, forcing T_2 OFF. Thus, the original stable state is now regained.

If the time taken for the stable-state to be regained is equal to t_d, then, this time t_d can be evaluated as follows. The equation for the voltage built-up across the capacitor is given by

$$V_c = V_{(final)} - \{V_{(final)} - V_{(initial)}\} \exp -(t/RC) \qquad \text{(i)}$$

where V_c = the voltage across the capacitor
= the voltage available at the base of transistor T_1 in our case, collector-end voltage being clamped at virtually zero level.

$V_{(initial)}$ = initial voltage across the capacitor at time $t = 0$.
= almost V_{CC} in our case.

$V_{(final)}$ = final voltage to which capacitor can charge in case it is allowed to charge indefinitely,
= V_{CC} in our case.

However, the capacitor can only charge upto a voltage equal to the cut-in voltage, hence, if we neglect this voltage, the above equation becomes

$$0 = V_{CC} - (V_{CC} + V_{CC}) \exp -(t_d / R_C)$$

This provides us with an equation for t_d as

$$t_d = 0.69\, RC \qquad \text{(b)}$$

It is clear that the resistance R provides a dual function. The first is to provide the base-current for transistor T_1, when it is ON, such that it remains in saturation in the stable-state of the circuit; and the second, to provide the reverse charging path for the capacitor, generating a time delay required of the circuit as a timer.

DESIGN OF MONOSTABLE MULTIVIBRATOR

The steps to be followed are similar, and in fact, are the same to some extent, as were followed for the design of bistable circuit. We shall repeat the steps here with required changes wherever necessary.

Step 1. Given supply voltage V_{CC}, and the biasing voltage V_{BB}, we can proceed to select the transistor to be employed from the required conditions.

Step 2. We select the operating current for the transistor from the conditions given. In the absence of the specified conditions, we select $I_{c(sat)}$ as about 5% to 10% of the $I_{c(max)}$ of the selected transistor.

Step 3. We calculate resistance R_c, as was done in design of BISTABLE circuit, by

$$R_c = \frac{V_{CC} - V_{CE(sat)}}{I_{c(sat)}}$$

and find the standard available value for the same. We recalculate the saturation current $I_{c(sat)}$. This value of the actual current will have to be used for further calculations.

Step 4. We determine the minimum value of base-current needed to support $I_{c(sat)}$, with the use of $h_{FE(min)}$. Assuming an O.D.F, we select a higher value for I_B.

Step 5. This current I_B is to be provided by resistance R.

$$I_B = \frac{V_{CC} - V_{BE(sat)}}{R}$$

It should be noted that, when transistor T_2 is on during its quasi-stable state, its base-current will have to be supplied by resistance R_1, as in the case of bistable multivibrator; in which case steps followed as in the case of bistable circuit will hold good.

Step 5A. This current I_{B2} is given by $(I_1 - I_2)$, where

$$I_1 = \frac{V_{CC} - V_{BE2(sat)}}{R_1}.$$

and

$$I_2 = \frac{V_{BE2(sat)} - (-V_{BB})}{R_2}$$

Step 6. Step 5A needs one more equation for finding values of the resistances R_1 and R_2. This can be obtained from the off condition of transistor T_2 during the stable-state of the circuit, since the base-emitter junction of this second transistor will be kept reverse biased by a negative voltage with the potential divider of R_1 R_2 combination. Hence, as was given in bistable circuit.

$$V_{BE2(off)} = \frac{R_2}{R_1 + R_2} \times (-V_{BB}) + \frac{R_1}{R_1 + R_2} \times V_{CE1(sat)}$$

If we assume that the $V_{CE(sat)}$ voltage is very small, and hence is negligible, then the second term of the above equation can be discarded, and the equation reduces to

$$V_{BE2(off)} = \frac{R_2}{R_1 + R_2} \times (-V_{BB})$$

Or,
$$R_1 = R_2 \times \frac{-V_{BB} - V_{BE(off)}}{V_{BE2(off)}} = n \times R_2$$

where $V_{BE2(OFF)}$ is negative.

Step 7. From the above two steps, we can now calculate the values of resistances R_1 and R_2.

Step 8. The only component that remains to be determined is capacitor C. For this purpose, time-delay required of the circuit should be given as data. We could make use of the equation $t_d = 0.69\,R\,C$, and calculate the value of the capacitor C.

One point of observation: Once we select a nominal value for the capacitor C as nearest **higher** value than the one calculated, we shall have to go back to the equation of the time-delay and recalculate a value for the resistance R. It may not be possible to have standard values of both the resistance R and capacitor C, which may satisfy the required condition. The solution would be to select a standard value of capacitor C and use a resistance R as a series combination of a fixed resistance and a potentiometer, which will also help us in getting a precise time-delay, in spite of tolerences of the components.

COMPUTER PROGRAM FOR DESIGN OF MONOSTABLE MULTIVIBRATOR

The computer program for the design of monostable multivibrator is given in Program P5. This program is similar to that for the design of bistable circuit.

This program also makes use of files for extracting data for the specified transistor and for finding a standard value of the resistance, either more than or less than the calculated value. After specifying the transistor number, the program finds the relevant data like maximum collector current, V_{ces}, the saturated transistor collector-emitter voltage, and the minimum gain, $h_{FE(min)}$. The program then calculates the resistance, R_c, and the resistance R. One can write a file for finding the standard value of capacitor C, once it is calculated from the equation of the time-delay t_d. However, this program assumes that the standard values of the capacitors are like those of the resistances, except that the values that we extract from the file should be labelled nF, nano-Farads. Calculation of the values of the resistances R_1 and R_2 is performed as was done for previous circuits. Calculation for finding the value of the potentiometer to be used for finer adjustments is done in exactly the same way as resistance R, using the same file, and adopting the same algorithm, assuming that the potentiometers are available in the same standard values as are fixed resistances.

Program - P5

```
PROGRAM monostable_multi (input,tranfile,resfile,output);
(* THIS PROGRAM DESIGNS A MONOSTABLE MULTIVIBRATOR *)
CONST
      odf   =  1.5;
      Vbes     =0.9;
      Vb1   =  -1.5;

VAR
      icm,Vces,hfm,rp,rn,ics,rc,ibm,iba,n,r1,r2      : real;
      Vb2,V1,Vc2,i2,i1,Vc1,iL,rL,td,R,C,Rpot         : real;
      Vop,Vbb,Vcc,i,e                                : integer;
      tranfile,resfile                               : text;
      tno,tno1,icm1,hfm1,Vces1                       : string[6];
      tranline                                       : string[30];

PROCEDURE resistor(ro:real; var rp,rn : real);

(* TO FIND THE STANDARD RESISTOR VALUE *)

VAR         d  :integer;
      r       :  real;
      rp1,rn1 :  string[3];
      resline :  string[7];

  BEGIN
   d := 0;
   WHILE ro >= 10 DO
    BEGIN
     ro   :=  ro/10;
     d    :=  d + 1;
    END;
   reset(resfile);
   readln(resfile,resline);
   rpl := copy(resline,5,3);
   val(rpl,rp,e);
   readln(resfile,resline);
   rnl := copy(resline,5,3);
   val(rn1,rn,e);
   WHILE rn < ro DO
    BEGIN
     rp := rn;
```

```
    readln(resfile,resline);
    rn1 := copy(resline,5,3);
    val(rn1,rn,e);
   END;
  rp := rp * exp(d * ln(10));
  rn := rn * exp(d * ln(10));
 END; (* end of resistor *)


BEGIN (* main *)
 ASSIGN(tranfile,'tran.pas');
 ASSIGN(resfile,'res.pas');
 writeln('MONOSTABLE MULTIVIBRATOR DESIGN ');
 writeln('------------------------------ ');
 writeln;
 write('INPUT VALUES OF Vcc AND Vbb    :    ');
 read(Vcc);
 write('          ');
 readln(Vbb);
 writeln;
 write('INPUT THE VALUE OF TIME PERIOD IN SEC.:');
 readln(td);
 writeln;
 write('INPUT TRANSISTOR NUMBER        :    ');
 readln(tno);
 rp := 0;
 rn := 0;
 writeln;
 writeln;


 BEGIN          (* accessing transistor file *)
 reset(tranfile);
  WHILE (tno <> tno1) DO
   BEGIN
    readln(tranfile,tranline);
    tno1 := copy(tranline,5,6);
   END;
  writeln;
  writeln('TRANSISTOR SELECTED FROM DATASHEET IS:',tno);
  writeln;
  icm1 := copy(tranline,12,3);
  val(icm1,icm,e);
  writeln('Ic(max) OF THE TRANSISTOR IS     :    ',icm:4:2,'mA');
```

```
    writeln;
    Vces1 := copy(tranline,16,4);
    val(Vces1,Vces,e);
    writeln('Vce(sat) OF THE TRANSISTOR IS     :    ',Vces:4:3,'V');
    writeln;
    hfm1 := copy(tranline,21,3);
    val(hfm1,hfm,e);
    writeln('THE VALUE OF hFE(min) IS          :    ',hfm:3:1);
    writeln;
    writeln('------------------------------------------------------');
    readln;
   END;

readln;

    BEGIN (* calculating value of Rc and R *)
     ics :=  0.1 * icm;
     rc  :=  (Vcc – Vces)/ics;
     writeln('CALCULATED VALUE OF Rc IS    :', rc:6:2, 'ohm');
     resistor(rc,rp,rn);
     writeln;
     writeln('STANDARD VALUE OF Rc IS  :    ',rn:6:2, 'kohm');
     rc  :=  rn;
     ics :=  (Vcc – Vces)/rn;
     writeln;
     writeln('RECALCULATED VALUE OF Ic(sat) IS:',ics:3:2,' mA');
     writeln;
     writeln('----------------------------------------------------');
     writeln;
     ibm :=  ics/hfm;
     iba :=  ibm * odf;
     R   :=  (Vcc – Vbes)/iba;
     writeln;
     writeln('CALCULATED VALUE OF R IS        :    ',R:6:2,' kohm');
    END;

    BEGIN(* calculating value of C *)
     C := td/(0.69*Rb);
     writeln;
     resistor(C,rp,rn);
     writeln('STANDARD VALUE OF C IS          :    ',rn:6:2,' nF');
     C   :=   rn;
     Rb  :=   (td/(0.69*C)) * le3;
```

```
  resistor(Rb,rp,rn);
  Rpot:=R – rp;
  writeln;
  writeln('STANDARD VALUE OF R IS          :   ',rp:6:2,' kohms');
  writeln;
  resistor(Rpot,rp,rn);
  writeln('STANDARD VALUE OF THE POT IS     :',rn:6:2,' kohms');
 END;

  writeln;
  writeln('------------------------------------------------------------');
  writeln;
  readln;
  clrscr;

 BEGIN(* calculating value of R1 and R2 *)
  ibm := ics/hfm;
  iba := odf * ibm;
  n   := (Vbb – Vb1)/Vb1;
  r1  := (n * (Vcc – Vbes) – Vbb – Vbes)/(n * iba);
  writeln;
  writeln('CALCULATED VALUE OF R1 IS   :', r1:6:2,' kohm');
  writeln;
  resistor(r1,rp,rn);
  writeln('STANDARD VALUE OF R1 IS  :   ',rp:6:2,' kohm');
  writeln;
  writeln;
  r2 := n * r1;
  r1 := rp;
  writeln('CALCULATED VALUE OF R2 IS   :',r2:6:2,' kohm');
  writeln;
  resistor(r2,rp,rn);
  writeln('STANDARD VALUE OF R2 IS  :   ',rn:6:2,' kohm');
  r2 := rn;
 END;
 writeln('----------------------------------------------------');
 writeln;
 writeln;

 BEGIN(* verification of the circuit *)
  Vb2 := (Vbb * r1)/(r1 + r2);
  writeln('VERIFICATION OF THE CIRCUIT ');
  writeln('--------------------------- ');
```

```
    writeln;
    writeln('BASE VOLTAGE OF OFF TRANSISTOR IS:',Vb2:4:2, 'Vo');
    Vc2 := (Vcc – ics*rc);
    writeln;
    writeln('COLLECTOR VOLTAGE OF ON TRANSISTOR IS:',Vc2:4:2,'
Vo');
   END;

    writeln;
    writeln;
    writeln('                    DESIGN COMPLETE.');
    writeln('                    ----------------');
    close(tranfile);
    close(resfile);
   END.
```

3.7 Design of Astable Multivibrator

In bistable circuits, both digital inverters are connected with each other by direct coupling. The collector of each of the transistors is connected to the base of the other transistor through resistance R_1. This resistance essentially supplies the base-current for the transistor which is ON, and also forms part of the network with R_2 to provide a fixed negative voltage to keep base-emitter junction of the other transistor to keep it OFF. As we have already seen, in a monostable circuit, one of the *dc* couplings is replaced by capacitor coupling. Capacitor C, replaces resistance R_1, connected between collector of transistor T_2 to the base of transistor T_1. To supply base-current to transistor T_1 when it is ON, a resistance R was connected between its base and the supply voltage V_{CC}. In astable circuit, both *dc* couplings are removed and are replaced be capacitors C_1 and C_2. The resistors R_1 and R_2 are used to supply the base-currents.

Refer to Fig. 3.11. The resistance R_1 and the capacitor C_1 decide the time for which the transistor T_1 is **OFF**, as in case of monostable circuit. This combination of R_1 and C_1 also forces the transistor T_1 from remaining OFF for a time more than ($0.69\ R_1C_1$), which generates the quasi-stable state for the monostable multivibrator. In the present circuit, we replaced even the second combination of R_1 and R_2 by R_2 and C_2. This, in turn, will prevent transistor T_2 to remain OFF for a time greater than given by ($0.69\ R_2C_2$). This circuit is, therefore, called an **ASTABLE** multivibrator, or a free-running multivibrator. There obviously is no need for external excitation for the transistor to change the state, as in the case of monostable multivibrator. None of the transistors remains either ON or OFF indefinitely, which generates a square waveform at the collector of each

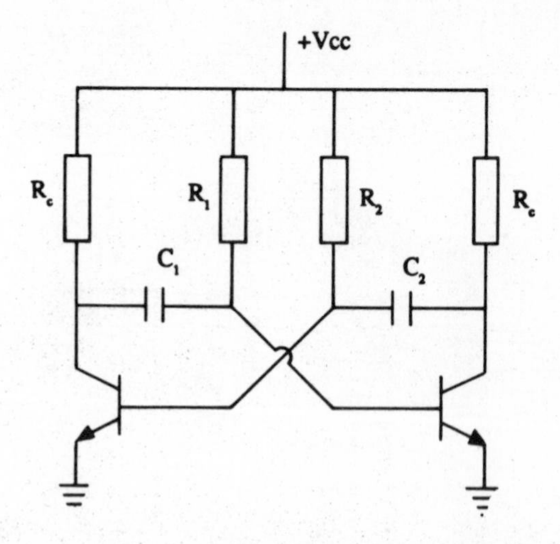

Fig. 3.11

transistor. The waveforms are complements of each other. If resistances $R_1 = R_2 = R$ (say), and also capacitors $C_1 = C_2 = C$ then the time t_1 (time for which the transistor T_1 remains OFF) is given by 0.69 R_1C_1, and the time t_2 (for which transistor T_2 remains Off or for which transistor T_1 remains ON) is given by 0.69 R_2C_2. Thus, the square wave will have a periodic time of

$$T = 0.69\, R_1C_1 + 0.69\, R_2C_2 \qquad \text{(i)}$$

$$= 1.38\ \text{RC}.$$

If, however, the square wave required at the collector, of say transistor T_1, has different ON and OFF times, like when we want to have ON time (low level of output voltage)), t_{on}, of 0.2 mSec (decided by R_2C_2 product), and OFF time (high level of output voltage), t_{off}, of 0.4 mSec (decided by R_1C_1 combination), the values of the capacitors, C_1 and C_2 will have to be varied accordingly. The resistors are normally kept equal, as they have to supply the base-current for the transistor concerned, and the transistors are assumed to be identical, having the same base-current requirements. The ratio of t_{on} to T is called a **duty cycle**.

DESIGN OF ASTABLE MULTIVIBRATOR

The steps to be followed are similar, and in fact are the same, to some extent, as were followed for the design of bistable and monostable circuits. We shall repeat the important steps.

Step 1. Given supply voltage V_{CC}, we can proceed to select the transistor to be employed from the required conditions. V_{CC} can also be determined from the waveform requirements, like its amplitude.

Step 2. We select the operating current for the transistor from the conditions given. In the absence of specified conditions, we select $I_{c(sat)}$ as about 5% to 10% of the $I_{c(max)}$ of the selected transistor.

Step 3. We calculate resistance R_c, as was done in design of BISTABLE and MONOSTABLE Circuit, by

$$R_c = \frac{V_{CC} - V_{CE(sat)}}{I_{c(sat)}}$$

and find the standard available value for the same. We recalculate the saturation current $I_{c(sat)}$, since resistance value is different from the calculated value. This value of the actual current will have to be used for further calculations.

Step 4. We determine the minimum value of base-current needed to support $I_{c(sat)}$, with the use of $h_{FE(min)}$. Assuming an O.D.F, we select a higher value for I_b.

Step 5. This current I_b is to be provided by resistance R.

$$I_b = \frac{V_{CC} - V_{BE(sat)}}{R}$$

Unless absolutely necessary, it will be highly convenient to use the same value of resistance for both the transistors.

Step 6. The components that remain to be determined are capacitors C_1 and C_2. For this purpose, periodic time required (i.e. the frequency of the waveform) should be given as data. We could make use of the equation $T = 1.38\ RC$, or $0.69\ R(C_1 + C_2)$, in case a duty cycle other than 50% is required, and calculate the required values for C_1 and C_2.

As in the case of monostable multivibrator, once we select a nominal value for the capacitors as nearest **higher** value than the calculated ones, we shall have to go back to the equation of the time-delay and recalculate a value for the resistance R. It may not be possible to have standard values of both the resistance R and capacitors C_1 and C_2 which satisfy the required conditions. The solution would be to select a standard value of the capacitors, and use resistances R as series combinations of fixed resistance and a potentiometer, which will also help us in getting a precision frequency in spite of tolerences of components used. Incidentally, since tolerences for commercially available capacitors are very poor, final changes may become necessary, most of which can be avoided, though, by judicious selection for the potentiometer.

COMPUTER PROGRAM FOR DESIGNING ASTABLE MULTIVIBRATOR

The program given in **P6** is written in Pascal for designing an astable multivibrator. Since this design is fairly simple, and is similar in some extent to a design of monostable multivibrator, the program written for monostable needs only a little modification for making it useful for designing astable multivibrator. The program given here, however, is slightly different in the sense that a different approach is made. Seemingly more difficult method of searching files for data for a given transistor is replaced by a straight forward method for asking the user to feed the data of whatever transistor he proposes to use. Though the method looks simple it presupposes that the user has the relevent data available with him when he desires to use the program for the design. Since the computer is being used for designing the circuit, it will be more appropriate for the data to be available in the computer memory (may be to make the computer look a bit more intelligent than it really is).
However, this program is written just to illustrate a different approach.

The program begins, as usual, with declarations of variables. Thereafter, two procedures are encountered. The first one is called **PROCEDURE RCLOSEST.**

Program P6

```
PROGRAM ASTABLE(INPUT,OUTPUT);
LABEL 10,20;
VAR
   ODF,VCESAT,VBESAT,ICM,HFEMIN,C1,C2,R1,R2,RA,RC,DUTY,FRE
Q,VCC:REAL;
   IC1,IB1,IBACT,T1,T2,T,D1,D2:REAL;
   POT:APRAY[1..7] OF REAL;
   TYPNO:STRING[10];
   L1:REAL;K1:INTEGER;
   PROCEDURE RCLOSEST(K:INTEGER;VAR RES:REAL);
   LABEL 10,20;
VAR
   R:ARRAY[1..14] OF REAL;
   N,J,I:INTEGER;M,P:REAL;
BEGIN
      WRITELN('ENTERED PROCEDURE1 RCLOSEST');
      IF RES>220000.0 THEN
                        BEGIN
                          writeln('RES is TOO LARGE');
                          GO TO 20; END,
     R[1]:=1;R[2]:=1.2;R[3]:=1.5;R[4]:=1.8;R[5]:=2.2;R[6]:=2.7;R[7]:=3.3;
     R[8]:=3.9;R[9]:=4.7R[10]:=5.6;R[11]:=6.8;R[12]:=8.2;R[13]:=10;
     N:=13; I:=1;
10:  FOR J:=2 TO N DO
  BEGIN
     M:=R[J-1]*I;P:=R[J]*I;
     CASE K OF
     0:IF ( M <= RES ) AND ( P > RES ) THEN BEGIN
                                    RES:=M;
                                    GOTO 20;END;
     1:IF ( M < RES ) AND ( P >=RES ) THEN BEGIN
                                      RES:=P;
                                      GOTO 20;END;
     END;
     END;
  I:=I*10;GOTO 10;
20:WRITELN('RES= ',RES:10:3) ;
END;
PROCEDURE CCLOSEST (VAR CP:REAL) ;
LABEL 10,20;
VAR
```

```
    J,N:INTEGER;
    I,M,P:REAL;
    C:ARRAY[0..20] OF REAL;
BEGIN
    WRITELN('ENTERED PROCEDURE CCLOSEST');
    IF CP>1000 THEN BEGIN WRITELN('***** ERROR ***** ');
    CP:=0.0;END;
    IF CP<1 THEN
          BEGIN
                C[0]:=0.0; C[1]:=1.0; C[2]:=1.2; C[3]:=1.5;C[4]:=1.8;
                C[5]:=2.0; C[6]:= 2.2; C[7]:=2.7;
                C[8]:=3.3; C[9]:=3.9; C[10]:=4.7; C[11]:=5.0;C[12]:=5.6;
                C[13]:=6.8; C[14]:=7.5;
                C[15]:=10.0;
                N:=15; I:=1.0E-06;
          END
        ELSE
        BEGIN
              C[0]:=0.0;C[1]:=1.0;C[2]:=1.5;C[3]:=2.0;C[4]:=3.0;C[5]:=4.7;
              c[6]:=50;
              C[7]:=6.0;
              C[8]:=8.0; C[9]:=10.0;
              I:=1.0; N=9;
        END;
 10: FOR J:=1 TO N DO
     BEGIN
          M:=C[J-1]*I;P:=C[J]*I;
          IF (M<=CP) AND (P>CP) THEN BEGIN
                                              CP:=P;GOTO 20;END;

          I:=I*10;GOTO 10;
          CP:=CP*1.0E-06;
          END;
 20:WRITELN('CP= ',CP*1E+06:10:3);
 END;
 BEGIN (*** MAIN BLOCK ***)
       WRITE('INPUT DATA:');
       WRITE('TYPNO: ');READLN(TYPNO);
       WRITE('VCESAT: ');READLN(VCESAT);
       WRITE('VBESAT: ');READLN(VBESAT);
       WRITE('ICM: ');READLN(ICM);
       WRITE('HFEMIN: ');READLN(HFEMIN);
       WRITE('VCC: ');READLN(VCC);
       WRITE('DUTY CYCLE IN % : ');READLN(DUTY);
       WRITELN('FREQUENCY IN HERTZ: ');READ(FREQ);
```

```
      IC1:=0.1*ICM;RC:=(VCC-VCESAT)/IC1;
      RCLOSEST(0,RC);
      IC1:=(VCC-VCESAT)/RC;WRITELN('IC1:= ',IC1:10:3);
      IB1:=IC1/HFEMIN;
      WRITELN('OVER DRIVE FACTOR: ');
      READ(ODF);WRITELN('ODF:= ',ODF:5:2);
      IBACT:=ODF*IB1;
      RA:=(VCC-VBESAT)/IBACT;
      RCLOSEST(0,RA);WRITELN('RA= ',RA/1000:10:3);
      IF RA=0.0 THEN GOTO 10;
      T:=1/FREQ;T1:=(DUTY/100)*T;
      T2:=T-T1;
      C1:=T1/(0.693*RA);C1:=C1*1.0E+06;CCLOSEST(C1);
      IF C1=0.0 THEN GOTO 10;
      C2:=T2/(0.693*RA);C2:=C2*1.0E+06;CCLOSEST(C2);
      IF C2=0.0 THEN GOTO 10;
      R1:=T1/(0.693*C1);RCLOSEST(0,R1);
      R2:=T2/(0.693*T2);RCLOSEST(1,R2);
      IF (R1=0.0) OR (R2=0.0) THEN GOTO 10;
      WRITELN('CHOOSE A POT ');
      D1:=RA-R1;D2:=RA-R2;
      POT[1]:=1.0;POT[2]:=2.2,POT[3]:=4.7;POT[4]:=10;POT[5]:=22;
      FOR K1:=1 TO 5 DO
   BEGIN
         L1:=POT[K1]*1000.0;
         IF D1<L1 THEN D1:=POT[K1];
         IF D2<L1 THEN D2:=POT[K1];
END;
      WRITELN('ASTABLE MULTIVIBRATOR DESIGNED USING ');
      WRITELN('TRANSISTOR TYPE:  ',TYPNO);
      WRITELN('VCESAT:  ',VCESAT:10:3);
      WRITELN('VBESAT:  ',VBESAT:10:3);
      WRITELN('ICMAX: ',ICM:10:5);
      WRITELN('HFEMIN: ',HFEMIN:7:2);
      WRITELN('DUTY CYCLE IN % : ',DUTY:7:2);
      WRITELN('FREQUENCY IN HERTZ: ',FREQ:10:3);
      WRITELN('THE DESIGNED VALUES ARE '):
      WRITELN('R1=  ',R1/1000:10:3,'  KILO  OHMS      WITH  A  POT
OF',D1:10);
      WRITELN('R2=  ',R2/1000:10:3,'  KILO  OHMS      WITH  A  POT
OF',D2:10);
      WRITELN('C1= ',C1*1E+06:10:3,' MICRO FARADS');
      WRITELN('C2= ',C2*1E+06:10:3,' MICRO FARADS');
10:END.
```

In order to provide a check for some unreasonable design, this procedure first checks whether the value of resistance calculated is more than 220 K. If it is, it will skip the procedure, tell the user that the "resistance is too large". If the calculated value is reasonable, then it finds two standard values of the resistances, one value nearest higher than the calculated, and the other nearest **lower** than the calculated value. Which value is to be selected is specified at appropriate places in the main program. This is done in the main program by assigning a value equal to 0 to a variable *k* if lower standard value of the resistance is needed, and assigning a value equal to 1 to *k* if higher standard value is needed. This is done by passing values to the parameter when the procedure **rclosest** is called.

The next procedure written is, likewise, for finding the closest value of the capacitor. Here also, if the value is unreasonably large, the procedure prints ******* error *******. In normal case, the procedure finds the value of the standard capacitor closest to the calculated value.

The standard value selection for the potentiometer is done exactly like that for fixed resistance. In fact, the same procedure, namely **rclosest**, could have been used for the potentiometer selection. However, in this program a separate block is written in the main program for this selection.

The last block of the program just prints out the values calculated for all active and passive components for the designed circuit.

3.8 Computer Program for Design of Amplifiers

Now since the strategy is clear, it will be easier to implement the design of amplifiers with computer aid. Design and analysis of computers were dealt with in Part-1 of this book. With the help of equations given therein, a program can easily be written. The following is the algorithm and steps followed for design of common-emitter amplifier.

Step 1: We feed the values for voltage gain required; Av, the stability S, output voltage magnitude required V_{op}, (rms value), supply voltage V_{cc}, the frequency of operation, and transistor parameters, like hie, hoe, hre, h_f(typ), hf(min).

Step 2: We calculate $R_c = Av \cdot \dfrac{hie}{h_f(\text{min}) - h \cdot Av}$

where $h = hie \cdot hoe - h_f(\text{typ}) - h_f(\text{min})$. We select higher standard value than the one calculated.

Step 3: Since, $V_{peak} = \sqrt{2} \times V_{op}$, we can select $V_{cc1} = 2 \times V_{CEQ} + V_{RE}$ where V_{RE} and V_{CEQ} can be assumed, respectively, as 10% of V_{cc} and $\dfrac{V_{CC}}{2}$, or may be fed as data to the program. *If* V_{cc1} calculated is greater than the given value of V_{cc}, then we take $V_{cc} = V_{cc1}$. Otherwise we, ignore V_{cc1}.

Step 4: Next we calculate $V_{ceQ} = \frac{V_{CC}}{2} - 1$ where 1 volt is to take care of saturation voltage.

Step 5: Since $I_c = \frac{V_{cc} - V_{CEQ} - V_{RE}}{R_c}$ we can calculate

$$R_E = \frac{V_{RE}}{I_c}.$$

We select standard (lower) value from the file.

Step 6: Normally $X_{CE} = \frac{R_E}{10}$,

Giving $$C_E = \frac{R_E}{10 \times 2 \times \pi \times f \times x_{cE}}$$

$$C_E = C_E \times 10^9\, nF.$$

We find C_E from the resistance file by the same procedure, assuming that condensor standard values are numerically identical to those of resistances.

Step 7: We refer to the equation for stability, which gives

$$S = 1 + \frac{1 + h_{fe(\text{typ})}}{1 + \frac{h_{fe(\text{typ})} \cdot RE}{R_B + R_E}}$$

whence, if $K = \frac{R_B}{R_E}$,

$$K = \frac{h_{fe(\text{typ})} \cdot S - h_{fe(\text{typ})} + S - 1}{1 + h_{f(\text{typ})} - S}$$

where $$R_B = \frac{R_1 R_2}{R_1 + R_2} \tag{i}$$

also, $$V_{cc} \cdot \frac{R_2}{R_1 + R_2} = V_{RE} + V_{BES} \tag{ii}$$

Step 8: We assume $X_{cc} = R_{o1} + R_{i2}$ in case two amplifiers are coupled together. From which C_c can be evaluated.

Program P7

```
program ceampr(input,output,mark);
(*this program designs ce amp.*)
const vces=0.3;
      vbes=0.7;
      vre=2;
var  vop,vcc,f:integer;
    ic,re,xce,ce,k,rb,vb,r1,r2,cc,ri2,xcc,ro1,av,rc,ro,vcc1,vceq,hre,hft,hoe,
    hfm,hie,h,v,vp,rp,rn,s:real;
procedure resistor(ro:real;var rp, rn:real);
(*this procedure selects std.resis. values*)
var d:integer;
    mark:file of real;
         r:real;
begin
    d:=0;
    while ro>=10 do
                begin
                   ro:=ro/10;
                   d:=d+1;
                end;
                assign(mark,'mark.pas');
                reset(mark);read(mark,r);
                rp:=r;read(mark,r);rn:=r;
                while r<ro
                do
                       begin
                          read(mark,r);
                          rp:=rn;
                          rn:=r;
                       end;
                 rp:=rp*exp(d*ln(10));
                 rn:=rn*exp(d*ln(10));
close(mark);
end;
begin(main program)
    writeln('input av,s,vop,vcc,f');
    readln(av,s,vop,vcc,f);
    (*to find rc*)
    begin
        writeln('hie,hoe,hre,hft,hfm');
        readln(hie,hoe,hre,hft,hfm);
        h:=hie*hoe-hft*hre;
```

```
    rc:=av*hie/(hfm-av*h);
    writeln('calc.value of rc=' ,rc:6:1,'ohms' );
    resistor(rc,rp,rn);
    writeln('std.value of rc=' ,rn:6:2,'ohms');
    rc:=rn;
end;
(*selection of vceq*)
begin
    vp:=vop*sqrt(2.0);
    vceq:=vp+vces;
    vcc1:=2*vceq+vre;
    if round(vcc1)>vcc then vcc:=round(vcc1);
    vceq:=vcc/2-1;
end;
(*to find re*)
begin
    ic:=(vcc-vceq-vre)/rc;
    re:=vre/ic;
    writeln('calc.value of re=' ,re:6:2,'ohms');
    resistor(re,rp,rn);
    re:=rp;
    writeln('std.value of re=',rp:6:2,'ohms');
end;
begin
    {to find ce}
    xce:=re/10;
    ce:=1/(2*3.14*f*xce);
    ce:=ce*1e9;
    writeln('calc.value of ce=',ce:6:3,'nf');
    resistor(ce,rp,rn);
    writeln('std.value of ce=',rn:6:3,'nf');
end;
    (*to find r1&r2*);
begin
    k:=(hft*s-hft+s-1)/(1+hft-s);
    rb:=k*re;
    vb:vre+vbes;
    r1:=vcc*rb/vb;
    r2:=(vb*r1)/(vcc-vb);
    writeln('calc.value of r1=',r1:6:2,'ohms');
    resistor(r1,rp,rn);
    writeln('std.value of r1=',r1:6:2,'ohms');
    writeln('calc.value of r2=',r2:6:2,'ohms');
    r1:=rn;
```

```
        resistor(r2,rp,rn);
        writeln('std.value of r2=',rn:6:2,'ohms');
        r2:=rn;
    end;
        (*to find cc*)
    begin
        ro1:=rc;
        ri2:=(rb*hie)/(rb+hie);
        xcc:=ro1+ri2;
        cc:=1/(2*3.14*f*xcc);
        cc:=cc*1e9:
        writeln('calc.value of cc=',cc:6:3,'nf');
        resistor(cc,rp,rn);
        writeln('std.value of cc=',rn:6:2,'nf');
    end;
    begin(* verification *)
    av:=hfm*rc/(hie+h*rc);
    writeln('the verified value of gain is=',av);
    rb:=r1*r2/(r1+r2);
    s:=(1+hfm)/(1+hfm*(1/(1+rb/re)));
    writeln('the verified value of stability=',s);
    end;
end.
```

Common Collector Amplifier Design:

1. Input : V_{opp}, *i/p* impedance, *o/p* imp, freq, $i_{c(max)}$, h_{fe}, h_{ie}

3. $V_{RE} = V_{opp}/2 + 0.01$
   $V_{EE} = V_{RE}$
   $V_{CC} = V_{RE} + V_{CE(sat)} + V_{CE}$

4. *To determine* R_b:
   $I_c = 0.05 * I_{c(max)}$
   $I_b = I_c/nfe$
   $V_B = V_{RE} + 0.7$
   $R_B = (V_{cc} - V_b)/i_b$
   Standardise R_b to closest previous value.

5. *To determine* R_E:

$$R_E = \frac{R_{out} \times \frac{(h_{ie} + R_b)}{(1 + hfE)}}{\left(\frac{h_{ie} + R_b}{1 + hfe}\right) - R_{out}}$$

   Standardise R_E to nearest higher value.

6. *To determine* C_c:
   If value of input impedance of next stage is given then:

$$R = z_{in} + R_E$$

$$C_c = \frac{1}{2\pi f R}$$

   If not given, then we assume $C_c = 0.1\mu F$.
   Standardise C_c.

2. Verify the circuit designed by the following:

$$R_{in} = R_b + [(1 + hfe)*R_E] + hie$$

$$R_{out} = R_E * \frac{(R_b + hie)}{(1 + hfe)} / R_E + \left(\frac{R_b + hie}{1 + hfe}\right)$$

$$A_v = R_E/(R_E + [hie/(1 + hfe)]$$

Program P8

```
PROGRAM coco(input,output,mark);
label 10;
const
    Vcesat = 0.4;
VAR
    Vcc,Vre,Vce,av,Vopp,f,rin1,rin,rout,hfe,hie,re,
    rb,cin,cc,icm,icq,rp,rn,Vb,ib,r,n,zin      : real;
                                                : file of real;
PROCEDURE resistor(ro : real; var rp, rn : real);
(* to find std values of resistances *)
VAR
    d : integer;
    r : real;
BEGIN
 d := 0;
 WHILE ro >= 10 DO
  BEGIN
   ro := ro/10;
   d := d + 1;
  END;
  RESET(mark);
  read(mark,r);
  rp := r;
  read(mark,r);
  rn := r;
  WHILE r<ro DO
  BEGIN
   read(mark,r);
   rp := rn;
   rn := r;
  END;
 rp := rp * exp(d * ln(10));
 rn := rn * exp(d * ln(10));
END;

BEGIN{main}
 ASSIGN(mark,'mark.pas');
 rp := 0;
 rn := 0;
 writeln('INPUT Vo/p-p, I/0 & 0/P imped, freq : ');
 readln(Vopp,rin,rout,f);
 writeln('INPUT hfe, icmax & hie of BC147A : ');
```

```
readln(hfe,icm,hie);

BEGIN {to determine Vcc}
 Vre := Vopp/2 + 0.01;
 Vce := Vre;
 Vcc := Vre + Vcesat + Vce;
 Vcc := round(Vcc);
 writeln('Value of Vcc =',Vcc);
END;

BEGIN {to determine rb}
 icq := 0.05 * icm;
 ib := icq/hfe;
 Vb := Vre + 0.7;
 rb := (Vcc – Vb) / ib;
 writeln('Calculated value of rb =',rb:6:2,'ohms');
 resistor(rb,rp,rn);
 writeln('Std. value of rb = ',rp:6:2,'ohms');
 rb := rp;
END;

BEGIN {to determine re}
n := (hie + rb) / (1 + hfe);
re := rout * n/(n – rout);
writeln('Calculated value of re = ', re:6:2, 'ohms');
resistor(rb,rp,rn);
writeln('Std. value of rb = 'rn:6:2,'ohms');
re := rn;
END;

BEGIN {to determine Cc}
  writeln('INPUT VALUES OF I/P imped OF NEXT STAGE; IF');
  writeln('NOT, TYPE 0');
  readln(zin);
  IF zin = 0
   THEN cc := 0.1 * 1e-6
   ELSE
    BEGIN
     r := zin + re;
     cc := 1 / (2 * 3.14 * f * r);
    END;
  cc := cc * 1e9;
  writeln('Calculated value of Cc =', cc:6:2,'nf');
```

```
  resistor(cc,rp,rn);
  writeln('Std. value of Cc =',rn:6:2,'nf');
 END;

 BEGIN {verifications}
  rin := (rb + (1 + hfe) * re + hie);
  writeln('Verified Value of I/P imped = ',rin:6:2,'ohms');
  n := (rb + hie) / (1 + hfe);
  rout := re * n/(re + n);
  writeln('Verified Value of O/P imped = ',rout:6:2,'ohms');
  av := re / (re + (hie/(1 + hfe)));
  writeln('Verified Value of Gain = ',av);
 END;
END.
```